COMPLETE GUIDE TO ELECTRICAL AND ELECTRONIC REPAIRS

COMPLETE

GUIDE TO

ELECTRICAL AND

ELECTRONIC REPAIRS

Carl G. Grolle

PARKER PUBLISHING COMPANY, INC.
WEST NYACK, N.Y.

Library of Congress Cataloging in Publication Data

Grolle, Carl G
   Complete guide to electrical and electronic repairs.

   Includes index.
   1. Electric engineering--Amateurs' manuals.
2. Electronics--Amateurs' manuals. 3. Household
appliances, Electric--Maintenance and repair--Am-
ateurs' manuals. I. Title.
TK9901.G76   621.3   75-20166
ISBN 0-13-160069-9

To Carl, Kenneth, and John

What This Book Can Do for You

This book is written for the homeowner-handyman who wants to save time and money repairing electrical and electronic devices. Each year the number of electrical items increases in the average home. Sooner or later they are going to need repairs.

Much electrical and electronic apparatus can be repaired without having a master technician's knowledge and with this book you will be able to fix countless items by using a little know-how, common sense, and simple test equipment.

Most electrical apparatus breakdowns are not difficult to diagnose and repair. You will save money as well as have a real feeling of accomplishment, so before you call the repairman why not try to fix it yourself? You will be successful so many times.

Remember the last time you were in the middle of cutting the lawn and the power mower stopped? You knew it would take a week or more to have it fixed and the grass was growing like weeds. Now you need not panic—fix it yourself! Check chapter 4 in this book on small gasoline engines and you'll probably be cutting grass again in no time.

Or how about that football game you have been waiting to see on TV. The set goes black. How can you get a repairman on Sunday? Don't fret. Maybe you can fix it right now . . . yourself! There are many things you can do to get that TV operating again. Read the television repair hints in chapter 13 and you might not miss the game after all.

Wouldn't you like to fix all the broken, small appliances that are lying around the house? Read chapter 5 so you can repair them and let them work for you again. There probably isn't much wrong with most of them.

Did you ever wake up in the middle of a winter's night feeling like you were frozen? This could happen if sometime during the night your furnace broke down. What would you give to get the furnace going again? If you are prepared, you *can* help yourself, and it won't be long before heat will be coming out of the radiators again. See chapter 12.

You may not think of your automobile as an electrical system, but most engine failures can be traced to a defect of an electrical nature. Very often an electrical failure is catastrophic, leaving you stranded in the middle of nowhere . . . or in heavy traffic. Wouldn't you like to be able to make emergency auto repairs and be on your way again? You can. Chapter 10 shows that it's easier than you may think.

This book gives suggestions and methods to repair all the preceding electrical-electronic troubles plus many, many other kinds. You will be successful in repairing most electrical problems, and the first time you succeed this book will have been worthwhile.

Carl G. Grolle

CONTENTS

COMPLETE
GUIDE TO
ELECTRICAL AND
ELECTRONIC REPAIRS

1

WHAT YOU NEED TO KNOW ABOUT ELECTRICAL SAFETY

Working with electricity can be safe or it can be dangerous. If you understand the basic principles of electricity and follow a few simple safety precautions, you can work with electrical devices without fear of being shocked.

What is electrical shock? What causes some people to be shocked when others aren't? When does a shock change from harmless to lethal? These questions and others are answered in this chapter.

Electricity can be dangerous without causing a shock. High-current, low-voltage batteries (automobile batteries) seldom shock, but can cause injuries if not handled properly. Find out the dos and don'ts when working with these devices by reading this chapter.

After you have become familiar with this chapter, you will be able to work with electrical devices around the house with complete safety. It doesn't matter if you are fixing a bad switch for the hall light, trying to repair your television set, or finding out why your car's ignition is defective, electrical safety must have number one priority.

1-1.   Understanding Electrical Shock

Voltage and current are two basic units of electricity. Current is the flow of electricity and is measured in amperes. Voltage is the pressure that moves electricity and is measured in volts. This question is often asked, what is the most dangerous part of electricity, current or voltage? The answer: Current kills! In order for current to pass through the body and do damage to the muscles and nervous system, however, there must be an adequate voltage present to force the current to enter the body.

Once current passes through the central body it takes only a relatively small amount of amperes to do significant damage. The following list shows the effects current has on the human body.

> .001 amperes causes skin tingling
> .009 amperes freezes muscles
> .03 amperes causing breathing to become difficult
> .075 amperes stops breathing
> .1 to .2 amperes causes ventricular
> fibrillation in the heart

Some people are more likely to be shocked than others when working with electricity. Individual skin resistance determines the severity of shock. Most people do have a comparatively high skin resistance, which prevents most shocks from being dangerous. Average dry skin resistance is about 100,000 ohms, but this resistance varies depending on the type of person and the dryness of the skin. Generally people who are chubby and perspire easily have the lowest skin resistance. On the other hand, a thin person with dry work-hardened hands would have the most resistance to electrical shock.

Current cannot enter the body due to the skin resistance unless the electrical pressure or voltage is strong enough to overcome the body resistance. The primary body resistance is the skin resistance. Underneath the skin the internal body resistance is relatively low in all persons. As soon as the voltage is strong enough, about 35 volts, current can be forced into the body in quantities strong enough to be felt. Figure 1-1 shows how the same voltage and current source could cause three different kinds of shocks in the same person.

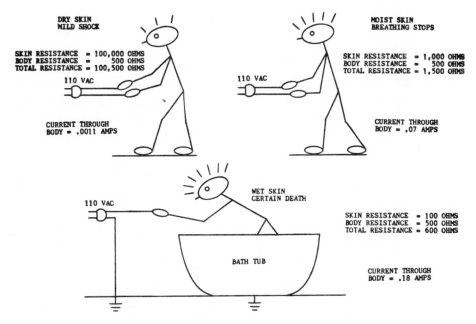

Figure 1-1.   Three kinds of electrical shocks from the same power source.

1-2. How to Work Safely with High-Current Batteries

A car battery is a good example of a high-current battery capable of producing over 100 amperes of current. Section 1-1 outlined how the heart can go into ventricular fibrillation when as little as .2 amperes passes through it. Yet people do not get shocked from handling a car battery. The majority of car batteries are 12 volts. This low voltage is not enough electrical pressure to force the tremendous current potential of the battery through the skin and into the body. For this reason, shock from an automobile storage battery is relatively unknown.

High-current batteries can be dangerous in other ways, however, and safety precautions must be used. Whenever fast charging a storage battery an explosive gas mixture forms beneath the cover of each cell. Any spark, match flame, or cigarette light could cause an explosion. Make certain you don't break live currents at the terminals of a charging battery because a spark can occur when the circuit is broken. Also when fast charging don't allow the battery to overheat (temperature exceeding 110° F). Overheating will damage a battery and cause the case to crack, spilling acid. In case you ever do come in contact with battery acid, make sure you flush the contact areas of your body immediately with large quantities of water.

Here's a way you can receive a nasty burn from a car battery and the reason why most professional auto mechanics don't wear rings. If you happen to be holding a wrench, screwdriver, or other metal object in your hand while resting on the battery, it is very possible for the wrench to accidentally short across the battery terminals and become sizzling hot almost instantly. If you are wearing a ring that completes the circuit with the tool across the battery posts, the ring will become red hot before you can move your hand (Figure 1-2).

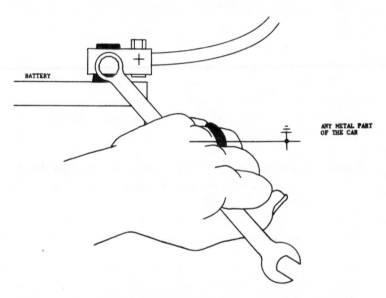

BATTERY

ANY METAL PART
OF THE CAR

Figure 1-2.   Accidentally short circuiting a storage battery while wearing a ring.

1-3.   Working Around House Wiring Circuits

Don't take chances of being shocked when working with house wiring cirucits. It only takes a few minutes to trip the circuit breaker or remove the fuse that deactivates the particular circuit with which you are working. After you have pulled the fuse, check the circuit with a voltmeter or test light to be absolutely certain no power is present.

House wiring circuits are generally color-coded with the hot wires being black and/or red; the ground or neutral wires are white. Both the hot and ground wires are current-carrying wires. Modern house wiring systems also have a third wire, which is the safety ground. Generally this wire is color-coded green and does not normally carry any current. This wire is used to make a common connection from all the metal junction boxes, conduit, and metal cases to the earth ground. Without the safety ground in a house wiring system, it would be possible for a hot wire to accidentally short circuit to a junction box or metal cabinet. Then anybody who happens to touch the case and is grounded would complete the circuit and receive a nasty shock. Most electrical codes demand that safety grounds be provided for any electrical apparatus that is adjacent to any grounded object (radiators, water pipes, cement floors, outside fixtures, etc.).

Figures 1-3A through 1-3F show various ways you can be shocked on house wiring circuits. Figure 1-3A points out that any time you are standing on a grounded surface and touch a hot line you can be shocked. Current will flow into your body through your hand and back to ground through your feet. Wearing a pair of thick, rubber-surfaced shoes would probably prevent this shock, but it is still an unsafe practice.

Figure 1-3B shows another way to become shocked due to an electrical circuit that has been installed improperly. The fuse has been

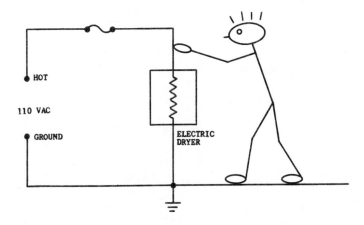

Figure 1-3A. Electrical shock from standing on a grounded surface and touching a hot line.

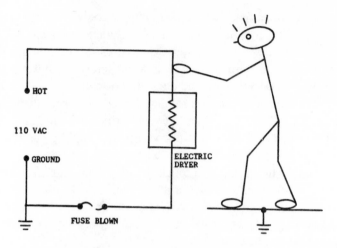

Figure 1-3B. Electrical shock from standing on a grounded surface and touching a hot line when the fuse has been installed improperly.

put in the ground line instead of the hot line. This time the fuse has blown due to a defect in the dryer, but shock still occurs because the ground line has been fused instead of the hot line.

In Figure 1-3C the heating element has shorted internally to the

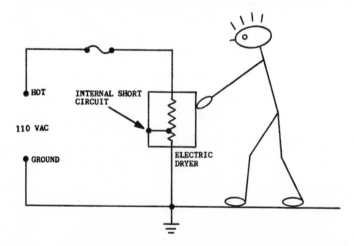

Figure 1-3C. Shock while standing on ground and touching the cabinet. The heating element has short circuited to the cabinet.

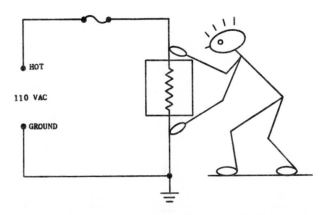

Figure 1-3D. Shock while standing on an insulated surface and touching the hot line with one hand and the ground line with the other.

dryer cabinet. Anyone standing on a grounded surface will be shocked when this cabinet is touched.

Any time you touch a grounded line with one hand and a hot line with the other you will be shocked. It doesn't matter whether your feet are insulated or on ground. Current will go in one hand and out the other passing through your chest in the process. See Figure 1-3D.

If you are working on a 220V circuit both lines are hot lines, Figure 1-3E. You will be shocked if you accidentally touch either hot line while standing on a grounded surface. Each hot line is 110V above ground.

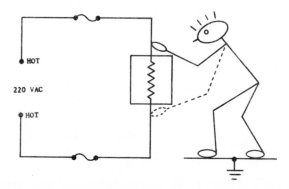

Figure 1-3E. Shock while standing on a grounded surface and touching either hot line in a 220V circuit.

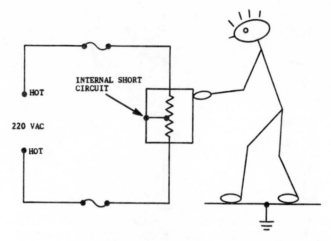

Figure 1-3F. Shock while standing on a grounded surface and touching the dryer cabinet. The heating element has short circuited to the cabinet.

Figure 1-3F shows how shock will occur if the dryer heating element accidentally shorts to the case. As soon as you touch any part of the cabinet while standing on a grounded surface, current will find a path to ground through your body.

Figures 1-3G through 1-3J show safe ways to work on and wire electrical circuits.

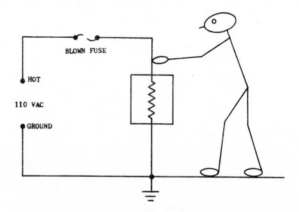

Figure 1-3G. Safe procedure for touching a hot line after the fuse has blown or has been removed even while standing on a grounded surface.

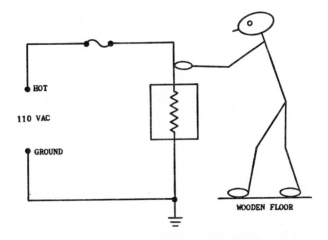

Figure 1-3H. Shock will not occur when standing on an insulated surface and touching a hot line even when the fuse is not blown. Just to be safe it would be wise to pull the fuse.

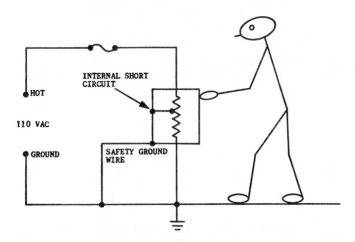

Figure 1-3I. Safe procedure for touching the dryer cabinet even when an internal short circuit has occurred as long as the case is fastened to ground with a grounding wire.

As long as the metal case, housing, or cabinet of an electrical device is grounded with a safety ground, the user is protected from shock if the device short circuits to its case. Most portable power tools

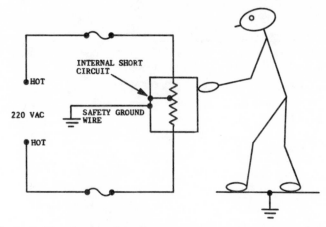

Figure 1-3J. Safe procedure for touching the dryer cabinet even when an internal short circuit has occurred as long as the case is fastened to ground with a grounding wire.

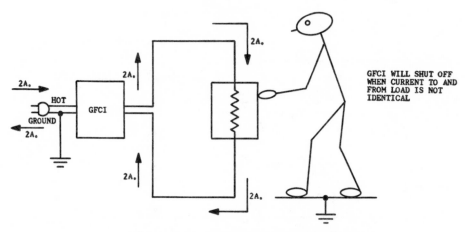

Figure 1-3K. Shock protection using a GFCI.

with a metal housing have a three-wire cord. The safety ground is color-coded green and is fastened directly to the metal housing of the tool. When the three-wire cord is plugged into a wall outlet make certain that the outlet is also equipped with the safety ground feature. This may be a separate green wire or the conduit housing, which is tied to earth ground.

A Ground Fault Circuit Interrupter (GFCI) is an electronic device that should be used whenever you are using a two-wire portable tool or on any electrical device when you are standing on a wet ground. Figure 1-3K shows how the GFCI protects you from a ground fault. If the tool is working correctly, exactly the same amount of current will flow through the hot and ground wires. If malfunction occurs in the tool shorting the case to the hot wire or if the user happens to come in contact with the hot wire, some of the current will pass through the person to ground. The GFCI will sense the lesser amount of current in the ground line and immediately shut off. The GFCI can sense a change in current as little as .005 amperes and shut off in less than .025 second. Any outdoor receptacles, especially those that are close to a swimming pool, should be protected wtih a GFCI.

1-4.   Methods for Working with High Voltage

This section deals with high-voltage, low-current apparatus used in an average house. Gasoline engines, television sets, and electronic air cleaners fit into this category. All these devices transform low voltage, high current into high voltage, low current.

If you touch a high-voltage wire with any part of your body, you will be shocked even if you are not grounded. There's a good chance you will be shocked even if you touch the insulation on a high-voltage wire. In fact, if the voltage is high enough, it will come out and meet you when you get too close; 20,000 volts will jump one inch through dry air. Most color television sets have a high voltage in the neighborhood of 25,000 volts. Any time you have your body within an inch and a quarter of the 25,000 V point, an arc will zap out and shock you.

When you are repairing or working with devices that use high voltage, make certain you know where the high-voltage points exist. Once power is applied, keep yourself and your tools a respectful distance from these areas. If you do have to work in these areas, make certain power is disconnected. Then take a test lead and fasten one end to the chassis or common ground point. With the free end of the test lead probe in the high-voltage circuit. This will discharge any high voltage that may have built up and stored in capacitors. See Figure 1-4.

Do not attempt to measure any high voltage with test instruments unless you are using specially designed high-voltage probes.

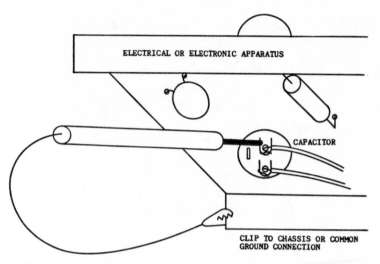

Figure 1-4. Using a test lead to discharge a high-voltage buildup.

A good safety procedure to use when working on devices where high voltage is present is to wear thick, rubber-soled shoes. You'll still get shocked if you touch a high-voltage point, but the shock will not be as severe. If you are working on a television set or other high-voltage apparatus, be sure to move it away from any radiators, water pipes, or any grounded objects. You'll be much safer if no grounded objects can be touched accidentally.

Follow these safety rules when working with high voltage:

1. Stay away from grounded objects.
2. Wear thick, rubber-soled shoes.
3. Identify the high-voltage points.
4. Discharge any high-voltage buildup on capacitors.
5. Don't touch or come close to any HV parts or leads when power is on.
6. Use only one hand to make tests while keeping the other hand in your pocket.

2

HOW TO USE
ELECTRICAL REPAIR EQUIPMENT

Most electrical-electronic defects need some type of test instrument to find or diagnose the trouble. A test light is the simplest of all troubleshooting tools. Read about how to use a neon test light in this chapter.

Probably the most useful and versatile of all electrical test instruments is the volt-ohm-milliammeter (VOM). The VOM is so functional because it is really four separate instruments in one: an AC voltmeter, a DC voltmeter, a current meter, and an ohmmeter. The use and application of this meter will be covered in this chapter.

This chapter also discusses the kinds of tools you will need for making electrical connections and stripping wire insulation. A few tools is all you will need to make a multitude of electrical-electronic repairs.

2-1. Ways of Using Neon Lamp Testers

Neon lamp testers are available in all sizes and shapes. Figure 2-1A shows a typical neon lamp tester. The tester consists of a neon

Figure 2-1A. Neon lamp tester.

lamp, a current limiting resistor, two probes, and some type of housing. Neon lamp testers are more versatile than incandescent lamp testers because neon bulbs have some particularly unique features. Most neon lamps need about fifty volts to light. Once they are lit they will glow brighter as more voltage is measured. So you can get a relative indication of the amount of voltage by the brightness of the bulb.

Another interesting feature of the neon bulb is its ability to distinguish between AC and DC voltage. When you are measuring AC voltage the gas around both electrodes will light, and when measuring DC voltage only the gas around the negative electrode will glow (Figure 2-1B). Thus, when the lamp glows you can tell whether the voltage is AC or DC, and if it is DC you will know which wire is positive and which is negative (polarity). Neon gas will also glow in the presence of a high AC energy field such as found in a television set or in the ignition system of an automobile.

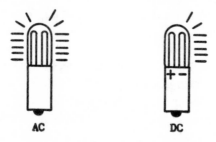

AC   DC

Figure 2-1B. Measuring AC and DC voltage with a neon lamp.

Figure 2-1C shows three ways to check a wall outlet with a neon lamp tester. If the lamp glows when the probles are inserted into each side of the outlet, it indicates that AC voltage is present. You can determine which side of the outlet is hot by touching one probe of the tester to ground (cold water pipe, etc.) and the other to one of the outlet

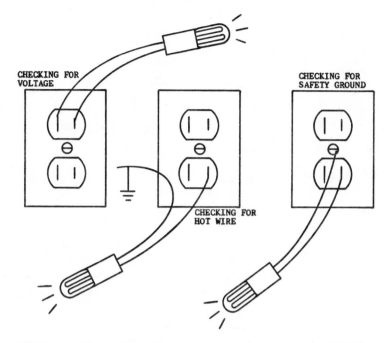

CHECKING FOR
VOLTAGE

CHECKING FOR
SAFETY GROUND

CHECKING FOR
HOT WIRE

Figure 2-1C. How to check a wall outlet with a neon lamp tester.

slots. The neon will glow when the hot side is touched. You can also determine if there is a safety ground (read section 1-3) in the outlet box by touching the tester probes between the hot side of the outlet and the metal junction box (metal screw that holds plastic front cover on). If the lamp lights, the box is grounded.

Here's a way to use the neon lamp tester to check out a typical 110 VAC electrical room heater (Figure 2-1D). As long as voltage is reaching all parts of the heater circuit the tester will glow. Start testing at position #1 to check the outlet. Then position #2 to check the cord; next position #3, etc. As soon as the neon does not light you have found the area of trouble. For instance, suppose neon #3 lit but not neon #4. This would indicate a thermostat that is defective or a defective wire between positions #3 and #4. To check the thermostat, place the neon across it as shown, #8 position. If the thermostat is okay the neon *won't* light. When the neon lights it indicates the thermostat switch is not closed and no power can reach the fan motor or heating element.

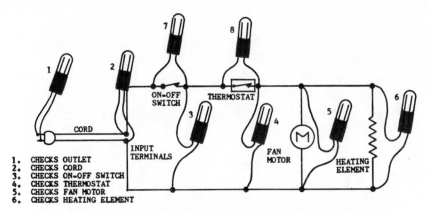

Figure 2-1D. Checking heater voltage with a neon lamp tester.

1. CHECKS OUTLET
2. CHECKS CORD
3. CHECKS ON-OFF SWITCH
4. CHECKS THERMOSTAT
5. CHECKS FAN MOTOR
6. CHECKS HEATING ELEMENT

When the neon tester lights in all positions except #7 and #8, either the motor or the heating element is defective. If the heating element is getting hot but the fan doesn't work, the fan motor is defective. Or vice-versa, if the motor is working but the heating element is not getting hot then the element is defective.

The neon lamp tester can also be used to detect the presence of high AC voltage. Just hold the lamp close to a high-voltage point and it will glow if high AC voltage is present. The probes should be kept in your hand with no connections to the circuit (Figure 2-1E). This tech-

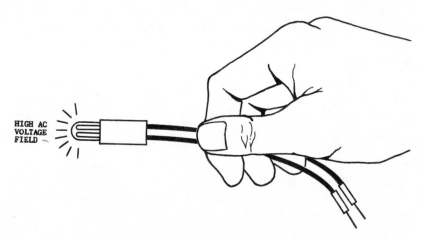

Figure 2-1E. Using a neon lamp tester to check for the presence of high AC voltage.

nique is very handy to use when you are working in a television set (chapter 13) or checking the ignition high voltage in an automobile (chapter 10).

Here are some test leads you can make that will really come in handy when you're using the neon tester or any time you're repairing or troubleshooting electrical devices. Figure 2-1F shows a selection of useful test leads. Sixteen-gauge lamp cord is great for making the leads. The wire is stranded to permit a lot of flexing yet it is heavy enough to carry adequate current for most applications.

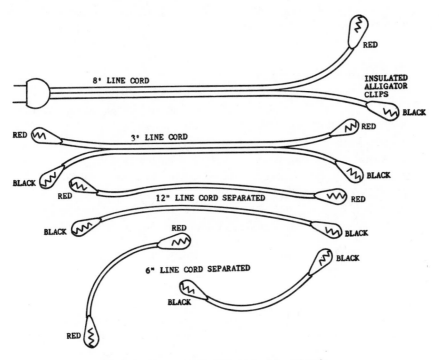

Figure 2-1F. Useful electrical test leads.

Make one eight-foot test lead with an AC plug at one end and a set of insulated alligator clips on the other. This is a real time saver when you need a source of line voltage. When you are using a test lead such as this, be sure to first fasten the alligator clips on to the terminals properly before inserting the plug into an outlet.

Cut a lamp cord three feet in length. Separate the two sides six inches on each end. Attach a red insulated alligator clip on each end of the same side. Attach a black insulated alligator clip on each end of the other side. Now you have two separate test leads electrically insulated but physically connected. A couple sets of leads such as this can certainly be helpful.

A few short alligator test leads also are needed. Separate a piece of twelve-inch line cord into two wires. Fasten a red insulated alligator clip on each end of one wire. Do the same with black insulated alligator clips on the other wire. Make a few test leads six inches long in the same fashion. You might want to terminate some of the test leads with mini-gator insulated clips for use with smaller apparatus.

2-2. How to Use a VOM

The volt-ohm-milliammeter is a necessity if you are going to be successful repairing electrical-electronic apparatus. You don't have to invest a lot of money for a good VOM, but make certain you have a meter that will be able to do the job. The following list shows some of the most important features a VOM should possess.

1. 20,000 ohms/volt DC sensitivity or more
2. 5,000 ohms/volt AC sensitivity or more
3. Overload protective circuitry
4. Adequate ranges for voltage, current, and resistance
5. Shockproof
6. Compact, hand-held, battery-operated.

Features 1 and 2 are important to make accurate measurements. Overload protective circuitry will protect the instrument when you accidentally make a measurement with the meter set on the wrong scale. Adequate ranges will allow you to make meaningful measurements over a wide range of voltage, current, and resistance. A shockproof meter will be able to survive a drop or other physical abuse. Feature 6 is strictly for convenience but is certainly worthwhile.

Probably the most confusing aspect of using the VOM is selecting the correct meter scale to read. There are always a multiple of scales on the meter face to take care of the various measurements, see Figure 2-2. Let's start with the top scale and work down. Resistance is read directly from the top scale in ohms. The function switch of the VOM

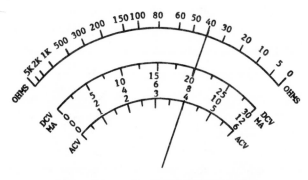

Figure 2-2.   Typical VOM meter face.

will be set on a resistance range R × 1, R × 100, R × 1,000, etc. When the function switch is on R × 1 the resistance measurement is exactly as indicated on the top scale. If the range switch is set on a range other than R × 1, simply multiply the reading by the appropriate multiplying factor. As an example, the meter resistance shown would be read as follows:

Resistance range	Ohmmeter reading
R × 1	40 ohms
R × 100	4,000 ohms
R × 1,000	40,000 ohms
R × 10,000	400,000 ohms

The second scale from the top is the DC voltage and current scale. Notice there are three sets of numbers directly beneath this scale that can be used. The numbers at the right of the scale indicate maximum range. For instance, if the function switch is set on 6 DCV, you would read the number scale that ends in 6; if the function switch is set on 30 DCV, you would read the number scale that ends in 30. The meter face shown in Figure 2-2 would indicate the following voltages:

DCV full scale	Voltage reading
.6	.4 volts
6	4 volts
30	20 volts
120	80 volts
600	400 volts
1,200	800 volts

DC current must also be measured using this same scale and numbers. Here's an example of what the meter would read when measuring current.

DC current full scale (MA)	Current reading
.06	.04 milliamps
12	8 milliamps
300	200 milliamps

The next scale down is the ACV scale. Use the same set of numbers you need for the DC ranges. The scale calibration is only slightly different for AC.

Most VOM meter faces follow this same type of pattern. Once you learn now to read one meter, you can read the face of any VOM in short order. Any particular problems can usually be cleared up by referring to the meter operating manual.

Here are a few pointers to always keep in mind when using a VOM.

For resistance measurements:

1. Short the leads together and zero adjust the pointer every time you change resistance scales.
2. Always measure resistance after power has been removed.
3. Do not touch the ohmmeter leads with your fingers while taking a measurement. Your body resistance will change the meter reading on high resistance settings.
4. The most accurate readings are made when pointer falls around the center of the scale.
5. When measuring for low resistance, keep the range switch on R × 1.
6. When measuring for high resistance, increase the range switch setting until a measurement can be made.
7. Most electrical parts' resistance can be measured accurately only when one of the part's leads has been disconnected from the circuit.

For voltage measurements:

1. Determine if the voltage is AC or DC.
2. Observe polarity if you are measuring DC. The black lead is the negative lead. The red lead the positive lead.

3. Always set the range switch on the highest setting and work down until you get a good reading.
4. Measure across the part or circuit for accurate readings.
5. Make certain your hands and body are insulated from the circuit and the meter probes.

For current measurements:

1. Most VOMs only measure DC current.
2. Observe polarity. DC electricity travels from negative to positive. The black probe (negative) should be tied to the most negative point that is being measured.
3. Always set the range switch on the highest setting and work down until you get a good reading. *Note:* It is not a good policy to change the range switch while measuring current. Stop power then change the range switch.
4. Current *must* be measured in series with the part or circuit. Cut or remove the wire and replace it with the VOM set on a current function.

2-3. Tools for Making Electrical Connections

Tools for making electrical connections can be roughly divided into two main groups: those that require melting solder and those that depend on a mechanical or crimping type of fastening method.

The handiest and most versatile tool for making connections while repairing electrical equipment is the soldering gun. The soldering gun will heat up in a few seconds and generally produce enough heat to solder even the largest connections. Don't scrimp when you buy a soldering gun because a good one will last for many years giving you excellent service over a wide range of soldering applications. A 200/275-watt dual heat soldering gun is great for general applications. Some guns have built in spotlights that really come in handy when soldering in a dark area. Here's a few tips for good solder joints:

1. Keep the tip clean and pit-free. File to shape when it becomes excessively corroded or pitted.
2. Heat the tip to operating temperature melting a coating of solder over its surface (tinning the tip). Do this before soldering any connection.

3. The joint or wires to be soldered must be clean and dirt-free.
4. Use 60/40 rosin-core solder.
5. Use rosin paste flux only when you are soldering old wires or connections that are impossible to clean thoroughly.
6. Keep maximum surface contact between the gun tip and the joint.
7. The solder will flow toward the gun tip; heat at the bottom of the connection. Apply solder to the top of the connection. This will force the solder to melt around and through the connection moving toward the tip.
8. Apply enough solder to give the joint a uniform coating of bright solder.
9. Don't allow the joint to move until the solder hardens.
10. Guns are made for intermittent duty. Don't keep the trigger pulled for long periods.

Most soldering guns are prone to oxidation between the heating element and the tip. A thin coating of oxides form where the tip is fastened to the gun. The gun tip will not heat hot enough when this occurs. To remedy this problem simply loosen and retighten the nuts or screws that hold the tip in place. This action will break the oxidation and the gun will operate like new again.

The soldering iron is not acceptable as a gun for repair jobs because it requires a relatively long time to heat. If you have to make a number of solder connections, the iron is the best tool. The iron is designed to be kept on for as long as necessary. Follow the same tips for using an iron as previously outlined for the gun. If you are using an iron, it's a good policy to use some type of holder or stand with it. This will keep you from being accidentally burned and help to prevent the iron from starting a fire from resting on something flammable.

There are many types of crimping tools available that use insulated or noninsulated metal fasteners to hold the wires together. This method is fast and convenient. The only drawback to using a crimping tool for making connections is that you need a supply of fasteners and lugs. Most crimping tools are designed for specific connectors. If you use the right crimping tool with the wrong connector, the results will probably be less than satisfactory. A crimped joint should not be able to be pulled apart. In fact the wires should break before the joint loosens. When in doubt as to the reliability of a crimped joint, you can always solder it after it is crimped.

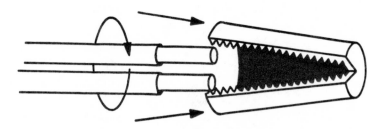

Figure 2-3. Cross section of a wire nut.

Wire nuts, Figure 2-3, make good connections when splicing wire. Choose the correct size nut to fit the wire gauge and the number of wires that are being spliced together. The wire ends are stripped and laid parallel to each other, then inserted into the wire nut. Screw the nut tightly on the wires until it is tight and the wire insulation is inside of the nut. Some wire nuts are made from plastic with internal threads; some have an internal spring wire to act as threads. If you are using the wire nut to make a connection in an area where temperature is a factor, be sure to use a ceramic-type wire nut.

2-4. Tools and Techniques for Stripping Wire

Whenever wire is going to be spliced or connections are made, the wire insulation must be removed. A good wire stripping will strip the proper amount of insulation leaving just enough wire to make the connection. The insulation should remain right up to the connection. In the process of removing the insulation the wire should not be cut or nicked. If you nick solid wire it will likely break at the nick as soon as the wire is flexed, moved, or vibrated. Cutting or nicking some of the wires in stranded wires will cause the rest of the strands to carry an excessive amount of current. Stranded wire is made to flex, so be sure to remove insulation from it very carefully.

The electrician's knife is the simplest tool to use for removing insulation from wires. When you cut wire insulation with a knife be sure to start the cut in parallel with the wire gradually tapering it into the wire. In this manner you will be able to feel when you are through the insulation and not seriously nick the wire (Figure 2-4A).

Figure 2-4B shows an inexpensive tool made especially for stripping wire insulation. A nut on the handle makes it adjustable for most wire gauges. The tool should be adjusted so that the inside opening is

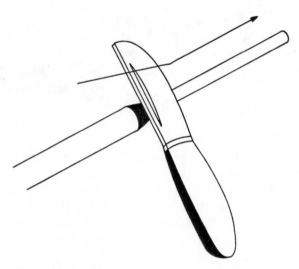

Figure 2-4A. Stripping wire insulation with a knife.

slightly larger than the wire size. Then when you cut through the insulation, the portion directly next to the wire will not be cut and certainly not the wire. After the insulation is cut around the wire simply pull on the tool. The remaining insulation will break free leaving a clean removal with no wire nicks. This tool requires a little practice to attain proficiency, but once the technique is mastered wire stripping becomes fast and easy.

Another type of stripping tool is shown in Figure 2-4C. This is a combination wire stripper and crimping tool. There is a series of sharp holes made in the inside of the handles. Each hole is designed to strip insulation from a specific gauge wire. Simply insert the wire in the

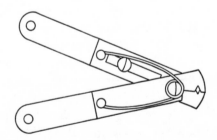

Figure 2-4B. Inexpensive wire-stripping tool.

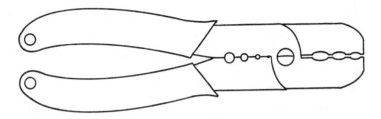

Figure 2-4C. Combination wire stripper and crimping tool.

correct hole and close the handles of the tool. Rotate the tool around the wire until the insulation is uniformly cut. Carefully pull on the tool until the insulation pulls free. Again, make certain not to nick the wire.

There are all sorts of other types of wire strippers around, including automatic ones. But for general use on electrical-electronic repairs around the house one of the previous kinds described should be sufficient.

2-5. Obtaining and Keeping Service Information

For any repair job if you have the manufacturer's service information available, you're going to save yourself a lot of time and difficulty. Many manufacturers include this data with the sale of their equipment. Some don't. You'll have to write to the company asking for the service information. Most companies will send it to you free of charge or just charge a nominal fee. Be sure to include the model number and/or other significant details identifying the item. If you don't have service information for equipment you are going to repair, obtain it as soon as possible. You won't want to wait when you need the information. The most useful information will include the following:

1. Parts list and parts numbers
2. Exploded view of parts
3. Electrical schematic
4. Pictorial wiring diagram
5. Timing diagram

The parts list will help you identify the replacement parts and ordering numbers. The exploded view will really come in handy when you're trying to figure out how to take something apart or put it back together again. Electrical problems become a lot more manageable when you can refer to a schematic. The pictorial wiring diagram is helpful to correlate the schematic with the physical parts. The timing diagram is necessary when you are checking out timed sequences such as found in an automatic washing machine or dishwasher. All kinds of service information can be helpful but the previous types are generally adequate for most repair jobs.

Gather your service information and organize it in a systematic manner and it will pay big dividends. Try this organizing method and you'll be able to put your fingers on the right information when you need it. Buy an inexpensive cardboard or metal one-drawer file cabinet. Organize it with alphabetical separators. Use it just for service information, nothing else. Gather up all the information you already have from all the various drawers, cupboards, hiding places, etc., and file it alphabetically. As soon as you obtain a new appliance or tool, file the service information immediately before it gets misplaced or lost. If you don't have the repair data, order it from the factory and file it away. Some day that information will save you a lot of trouble and expense.

3

EASY REPAIRS FOR
SWITCHES, FIXTURES, AND LAMPS

In this chapter you will find some ideas on how to repair all sorts of electrical problems found in lighting and switching devices. Common ceiling and wall fixtures occasionally become defective and need repairing or replacing. You'll find these repairs quite simple to make. The next time a floor lamp or table lamp stops working, don't think about replacing it . . . fix it. You'll be successful.

Fluorescent lights are slightly more complicated than ordinary incandescent lamps, but once you understand what makes them work, they, too, are quite easy to repair. Maybe you have a small fluorescent desk lamp that isn't working anymore. You really miss it. Don't fret! You can fix it yourself in short order. The same goes for the defective fluorescent fixture in the kitchen.

High-intensity lamps are great for placing a bright light in a small area. Once you become accustomed to a high-intensity lamp, you are going to miss it when it breaks. Sure, you can replace it with a new one, but why not fix it yourself for double satisfaction?

Do you know how many lamps can be found in an average American automobile? It's not unusual to find thirty-five or more. In this chapter you'll find out how you can simplify and repair all kinds of light problems in your car.

3-1. Repairing Switches, Ceiling and Wall Fixtures

Let's start with a single-bulb, ceiling-mounted light fixture that is turned on by a wall switch. It doesn't turn on. Whenever a trouble occurs, make a list of possible causes starting with the most likely suspects and ending with the least probable.

1. Bulb burned out
2. Fuse blown
3. Defective switch
4. Bad connections
5. Bad socket
6. Broken wire

Analyzing the list will show why the order of possible defects was chosen. Bulbs operate white hot and have a limited life span. A bulb should be the first thing to check. Next, check if other lights or outlets are working, but not necessarily in that same room. A fuse may have blown or a circuit breaker activated from an overload somewhere else. Occasionally a fuse can open up just with old age. Next, check the switch using a neon tester. Switches are mechanical devices that also have a limited number of operations. Then check for bad connections. When connections are not made tight, they can oxidize and burn, blocking the flow of electricity. The socket could become defective after a number of light bulbs have been screwed in and out or it could just have been manufactured poorly. Last, and most unlikely, a wire may have become broken. This very seldom happens unless the wire has been vibrated or flexed continually.

Here's how to check a single action wall switch with a neon tester (Figure 3-1A). Unscrew the wall plate being careful not to touch any electrical connections. Pull the switch out enough to make the test. If the switch is good the tester will light on terminals A and B when the switch is on. It will light on A or B when the switch is off but not in both positions. The side that lights is the hot wire carrying electricity to the switch. If the switch is bad, turn off the power at the main fuse box and replace it. This type of switch is the most common type used in house wiring circuits.

A more complicated switch is the three-way type shown in Figure 3-1B. When one light must be turned on and off from two different locations this type of switch must be used. Two three-way switches are

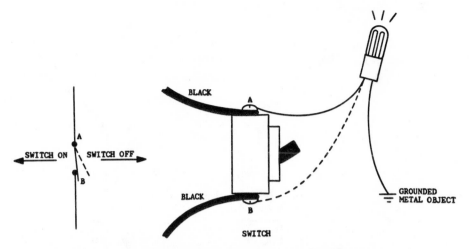

Figure 3-1A. Checking a single-action wall switch with a neon tester.

used in conjunction with the light (load) they are operating. Figure 3-1C shows how to check out these switches using a neon tester. Unscrew the wall plate for each switch being careful not to touch any of the connections. Pull out the switches so you can reach the terminals. Locate the main hot wire fastened to terminal A. This will be the only one of the six terminals that has power on it for any position of the switches. Then move the ungrounded tester probe to B; the tester

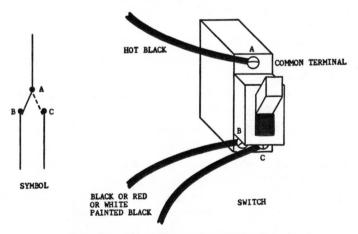

Figure 3-1B. Three-way switch wiring.

SWITCHES, FIXTURES, AND LAMPS / 43

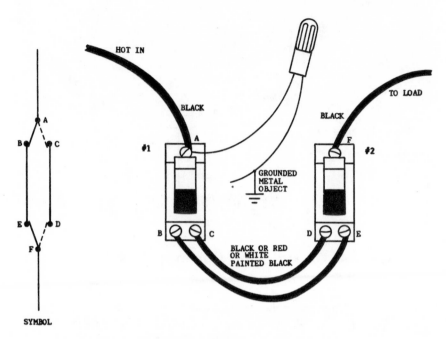

Figure 3-1C. Checking a three-way switch system with a neon tester.

should glow on and off as the switch #1 is activated. Probe terminal C and the same action should occur. C will light the tester when B won't and vice-versa. This checks out switch #1. Touch the ungrounded probe to D. The tester will light alternately as switch #1 is activated. Do the same for terminal E. If this checks out the wiring between switches is okay. Move the probe to terminal F. The tester should light alternately as switch #2 is activated regardless of the position of switch #1. Now switch #2 is checked. When you have to replace a defective three-way switch with a new one, be sure to identify the common terminal first. All three-way switches do not have the common terminal in the same physical position. *Note:* You can make an ohmmeter check to verify terminals on the replacement switch.

In most homes the three-way switch is the most complicated house wiring switch. However, there are some lights that can be switched on and off from more than two places. This type of circuit requires one four-way switch between a pair of three-way switches. Refer to Figure 3-1D for an explanation on how to check a four-way switch with a neon tester. The tester on terminal A should light alter-

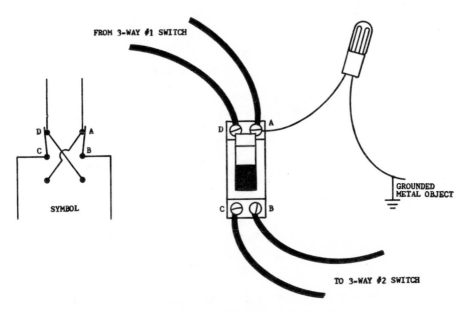

Figure 3-1D. Checking a four-way switch with a neon tester.

nately as the three-way switch #1 is switched. The same action should occur on terminal D. When A lights the tester, D won't, and vice-versa. The neon should light on B whenever it lights on A, and it should light on C whenever it lights on D for one four-way switch position. On the other four-way switch position A and C should light the tester as well as D and B, depending on the three-way position. If you have to replace a four-way switch, be sure to ohmmeter the contacts to identify them. The terminals are positioned differently for various brands.

Suppose you want to check out a ceiling fixture. Unscrew the fixture and pull it down until the wires are reachable. Most fixtures have black and white wires respectively connected to black and white wires in the junction box (Figure 3-1E). Connect the neon tester between the white and black wires coming into the junction box. Most connections are made with wire nuts. Usually there is room enough to slide the neon test probe into the bottom of the nut to make the connection. If the wall switch is on, the tester should glow. Next check each of the fixture sockets with the tester. Touch one probe to the inside bottom of the socket and the other to the metal-threaded portion. If the

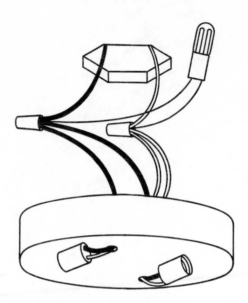

Figure 3-1E. Testing a ceiling fixture with a neon tester.

socket is good the tester will light. If the tester lit at the junction box but not at the socket, there's a broken wire or defective connection between the junction box splice and the socket. Turn off the power to the fixture and check continuity of the fixture leads with an ohmmeter. The black junction box wire should have zero ohms resistance to the inside center terminal of each socket. The white junction box wire should have zero ohms resistance to all of the metal socket threads.

Wall-mounted fixtures are basically wired the same as ceiling fixtures. One basic difference is that an always-on receptacle may be part of the fixture. This type requires two black wires and one white to operate if it is switched from a remote location. One black wire switches the fixture on and the other provides always-on power to the receptacle. To test this fixture, the neon should light when either black wire is probed while the other tester probe is on the white wire (Figure 3-1F).

3-2.   How to Fix Incandescent Lamps

Let's start with a single-brightness desk lamp. Most desk lamps have only four basic electrical parts, not counting the light bulb: plug,

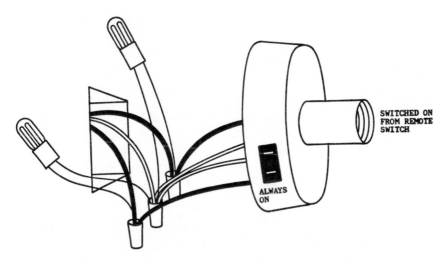

Figure 3-1F. Testing a wall fixture with a neon tester.

cord, switch, and socket. When the lamp becomes defective you can bet the trouble will be where the most physical use has occurred. If the cord has been plugged in and out repeatedly over the years, there is a good chance the wires will break inside the insulation close to where the cord enters the plug. You can usually confirm this by turning the switch on and wiggling the cord back and forth near the plug. If the lamp goes on when the cord is held in a certain position, there's a good chance you've found the culprit. To fix this problem simply cut off the plug and replace it with a new one. Figure 3-2A shows some of the ways to connect line cords to plugs. Plug A is the conventional re-placement plug using two screw connections. This is a good plug design because it gives you enough room on its body to grip when pulling it out of the outlet. It also can be used with all types of line cords. Separate the line cord about two inches by pulling the two sides apart. Strip the insulation off ¾ of an inch from each side. Tin the wires (lightly coat with solder) for about ¼ inch. Insert the prepared end of the line cord through the plug. If there is enough room tie a knot in the cord at the bottom of the separation. This will provide some strain relief if someone pulls the plug out by the cord. Look at the prongs closely before routing the wire to the connection screws. Most prongs are designed so that the line cord can be wrapped around them one-half turn on the way to the screw. This also provides strain relief. Wrap the wire completely around the screw in a clockwise direction so

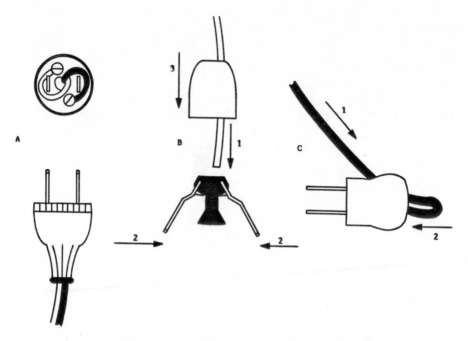

Figure 3-2A. Connecting line cords to plugs.

when you tighten the screw it will tend to pull the wire under its head. No strands of wire should be seen and the wire insulation should meet the screw head. After you have tightened the screws, don't forget to insert the insulating wafer that fits over the prongs. This type of plug is the most time-consuming to replace but also the most durable.

Plug B, Figure 3-2A, needs no special wire preparation. Grasp the plug body in one hand; with the other hand compress the prongs toward each other and pull. The prongs and inside part of the plug will pull out of the housing. Now push the prongs apart as far as they will move. You will see a sharp point on each side of the prongs. Cut the line cord evenly with no cord separation and insert through the hole in the housing into the hole in the prong assembly pushing the cord until it bottoms. Then squeeze the prongs back to their normal position. This will automatically puncture the insulation on each side of the line cord making the connection. Slide the housing back over the prong assembly until it snaps. That's it!

The third type of plug, C in Figure 3-2A, is also very quick to assemble. Cut the line cord square with no cord separation. Slide the prepared line cord end through the side plug hole so that it comes out of the top. Then grasp the cord and double it back straight down into the plug as far as it will go. You'll feel a little opposition as the sharp points of the prongs enter each side of the cord. When the cord bottoms the connections are made. Remove the excess loop from the top of the plug by pulling the remainder of the cord back through the side hole. It works like a charm.

If the plug on the desk lamp is okay, the next obvious part to check is the switch. When the light stays on regardless of the switch position, the switch is undoubtedly bad. If the light never comes on regardless of the switch position then the switch could be bad. Most lamps have the switch built directly into the socket so there's not much use in trying to find out if the trouble is in the switch or socket. If you replace one, you'll replace both.

Figure 3-2B shows an exploded view of a typical socket-switch assembly. The base is threaded and screws on a hollow, threaded pipe within the lamp housing. It is held in place by a locking screw on the base. The case is attached by a series of notched indentations that snap into similar indentations in the base. To separate the base from the case remove the on-off knob. Just turn the knob in the opposite direction it turns on (counterclockwise) and it will unscrew. Then thumb press on the case next to the slot. Most cases have "press" stamped in this area. A fair amount of pressure is necessary to unlock the case from the base. Once they unlock, all the parts with the exception of the base will pull out easily. The line cord will still be attached to the switch-socket assembly so help the line cord by pushing it in the bottom of the lamp while pulling from the top. You may have to remove the lamp bottom before the line cord can be pulled through the lamp stem. *Caution:* Be careful not to damage any cord insulation

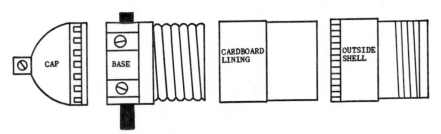

Figure 3-2B. Socket-switch assembly.

while pulling it through the lamp. Inspect the line cord very carefully for deterioration. At this stage it will only take a few more minutes to replace the line cord giving you a first-class repair job. Replace the switch-socket assembly by attaching the line cord in a manner similar to the connection to plug A, Figure 3-2A. Reassemble and try the lamp operation. It will work like new!

The other common type of incandescent lamp uses a three-way bulb. Many floor lamps and large table lamps are designed for this kind of operation. Figure 3-2C shows how the internal construction is made in a three-way light bulb. There are two filaments and three connections on the base of the bulb. The threads, center base terminal, and a base ring terminal make the connections to the socket. The four-position switch moves from off to 50W, to 100W, to 150W, and back to off again. Repair of a three-way lamp is very similar to that of the single-brightness lamp just described. When you buy a replacement socket-switch assembly, be sure it's designed for three-way operation.

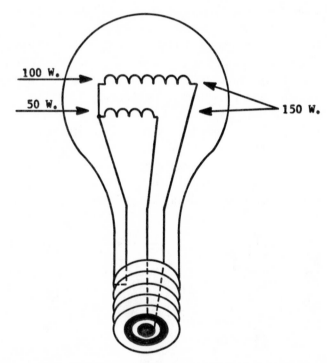

Figure 3-2C. Internal construction of a three-way light bulb.

Look into the inside of the socket. You'll see two contactor lugs at the bottom if it is a three-way socket.

3-3. Practical Methods for Repairing Fluorescent Lights

Fluorescent light fixtures can be divided roughly into two categories: one type that requires a starter to turn on and the other type that requires no starter. Figure 3-3A shows the circuit for starter operation. The switch, filament, starter, filament, and ballast are all connected in series. Any part that becomes defective will shut down lamp operation.

This circuit operates as follows. When the on-off switch is closed the starter will enable current flow through the series circuit, heating both filaments. The heat from the filaments will vaporize mercury inside of the tube causing mercury vapor to form. After a short time delay, the starter will automatically open, stopping current flow in the circuit. As soon as current stops flowing, a magnetic field collapses around the ballast causing current to jump from one filament through the mercury gas to the other filament and starting up the circuit again without any current flow in the starter. Current flow in the tube will cause the phosphor coating to fluoresce, producing the light. Now the ballast acts as a limiting device to protect the lamp from drawing excessive current. Each time the lamp is turned on, this series of events must happen to produce light.

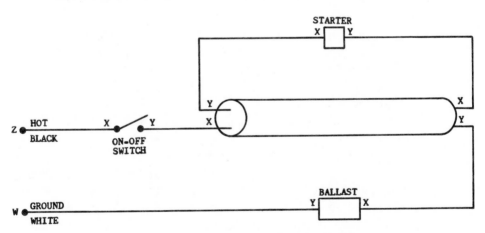

Figure 3-3A. Fluorescent light with a starter circuit.

When repairing fluorescent lights remember that parts are not interchangeable unless they are from identical units. Suspect trouble in the following order:

1. Fluorescent tube
2. Starter
3. On-off switch
4. Connections
5. Ballast

First try an identical new fluorescent tube. If it is a two-tube fixture substitute two new tubes. When the tubes are not faulty check the starter. If you have an identical starter handy just substitute. Or you can jump the starter terminals with a clip lead to simulate starter action. Removing the clip lead will start the light if the starter is defective.

You can troubleshoot the circuit using a neon tester but in the majority of cases substitution of the tubes and the starter will be all that is necessary. If you need the neon tester to find the problem in a dead circuit refer to Figure 3-3A. Put the tester leads on Z and W. The neon should light. Make certain the on-off switch is on for the following tests. Place the tester across each of the parts, one lead to X, one lead to Y. The neon should not light. When the tester lights you have found the trouble area. For example, if the neon lights when you are on the prongs of the tube, it means that the filaments on that end of the tube are bad or there is a bad connection on the socket. A good fluorescent tube filament will measure low resistance across its prongs.

Ballasts seldom become defective. You can check the continuity for a low resistance, but the only certain check is to replace it with an identical ballast. A common ballast problem is hum. All ballasts hum to a certain extent, but sometimes the ballast mounting hardware will become loose. If this is the case, tighten up the connections and most of the hum will disappear. When this won't do the trick, a new ballast is the only answer.

Some fluorescent desk lamps have a special manual start, on-off combination switch. This is actually two switches in one; a standard on-off switch and a momentary contact switch for the starter. Proper operation of this switch is easily checked with an ohmmeter.

Many fluorescent lamps do not need starters. They use special instant-start tubes and ballast. Troubleshoot these fixtures in much the same way as for the starter type. Figure 3-3B shows the circuit hookup for a typical instant-start fluorescent lamp.

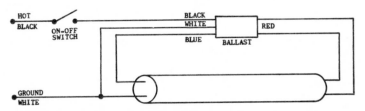

Figure 3-3B. Instant-start fluorescent lamp circuit.

3-4. Understanding High-Intensity Lamps

High-intensity lamps use low-wattage bulbs with a close-fitting metal shade that directs a lot of light into a small area. There are two types of high-intensity lamps: one type whose bulb operates directly from 110 VAC and one whose bulb operates on 12 VAC supplied by a step-down transformer built into the lamp base. The lamp voltage is marked on the bulb. The 110 VAC base has screw threads, while the 12 VAC bulb is twist-locked similar to automobile bulbs so they can't accidentally be interchanged.

To repair the 110 VAC high-intensity lamp, use just about the same techniques as you would for the conventional incandescent lamp as outlined in section 3-2. The socket is generally held to the lamp housing with a single screw. One precaution you should observe when replacing a socket is the heat factor. The socket and shade, due to their compact construction, are subject to a great deal more heat than most ordinary lamps. For this reason, make certain to use wire, insulation, and electrical tape made to withstand the higher temperatures. Very often the switch and socket are separate units so they can be replaced individually. You may have trouble finding an exact replacement switch since they are made in a variety of designs. However, you should be able to find an adequate substitute that has the same basic dimensions. Just be sure the electrical specifications are equal to or greater than the original.

Figure 3-4A shows the circuit of the transformer-type high-intensity lamp. Use the VOM (see section 2-2) to troubleshoot a defective lamp. Table 3-4 shows the correlation between the VOM voltage measurements and probable trouble.

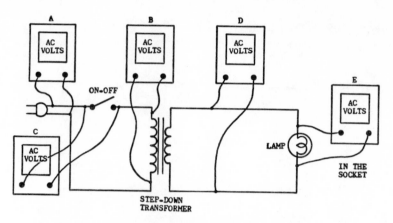

Figure 3-4A. Transformer-type high-intensity lamp circuit.

Sometimes an apparent defective transformer can be repaired (see section D, "Remedy," in Table 3-4). One of the most common failures is a broken transformer wire. These transformers have fine wires soldered to their terminals with no provision for strain relief. Any bumping, dropping, or other mechanical strain can snap these connections. This type of defect can often be repaired. Carefully peel back the insulation tape around the windings until the transformer terminals are

Table 3-4

VOM	READING	CAUSE	REMEDY
A	110 VAC	Normal	
	0 VAC	Bad plug or line cord	Replace
B	110 VAC	Normal	
	0 VAC	Defective switch	Replace
		Bad connection	Repair
C	110 VAC (Switch off)	Normal	
	110 VAC (Switch on)	Defective switch	Replace
D	12 VAC	Normal	
	0 VAC	Bad transformer	Replace
		or connection	Repair
E	12 VAC	Normal	
	0 VAC	Bad socket	Replace
		Bad connections	Repair

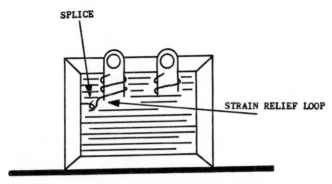

Figure 3-4B. Repairing an open transformer winding.

exposed. The chances are the wire will break where it is fastened to its terminal. When you see the break, remove the varnish insulation from the remaining wire with fine sandpaper, being very careful not to break any more of the wire. If the wire breaks again just as it comes out of the winding, you won't be able to repair it. Splice another wire on the broken lead, leave a loop for strain relief, Figure 3-4B, and resolder it on the terminal. Retape, test and the transformer probably will operate like new!

Figure 3-4C shows how to check the resistance of the transformer

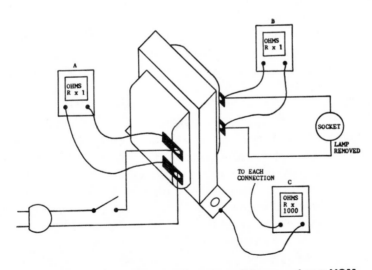

Figure 3-4C. Checking transformer resistance using a VOM.

using a VOM. Ohmmeter A will measure the primary resistance, which should usually be less than one hundred ohms. Ohmmeter B will measure the secondary resistance, which will be lower than the primary. This resistance will measure only a few ohms. Ohmmeter C will measure the resistance from the transformer case to the windings. This measurement should be too high to detect with the VOM.

3-5. Guide to Servicing the Automobile Lighting System

All lamps in the automobile require only one wire per filament. The other wire is the frame of the car. Lamp voltage is the same as the battery: 12V lamps for 12V batteries, 6V lamps for 6V batteries. You can troubleshoot automobile lighting problems with a VOM or you can make a light tester using a parking lamp bulb with some test leads soldered to the connections.

Figure 3-5A shows a simplified typical wiring layout of the outside lights for an automobile. The headlights are protected with a circuit breaker, which is generally built into the headlight switch. This is to insure that the headlights will not fail due to an intermittent short circuit. The remainder of the outside lights are protected with a fuse. Whenever you have a lighting problem, carefully analyze the symptoms before you do any testing. The majority of problems can be isolated just by referring to the lighting diagram using logical reasoning and common sense.

Let's discuss some typical problems that you may have and how to go about the repair. Suppose the right rear tail lamp is not lighting. First, check the rest of the outside lights for proper operation including the brake lights. In this case everything seems to be okay and the right rear tail lamp does light when the brakes are applied. The majority of lighting defects are caused by a burned-out light bulb and this is a typical case. The tail lights are dual filament lamps. One filament is used for the tail light while the other is used for the brighter brake and turning signal light. The logical conclusion is that the tail light filament has become defective. Replacing the lamp with the same number bulb cures the trouble. To remove a tail lamp, simply push in on the bulb and turn about one-quarter turn counterclockwise. It will pop out.

Now let's take the same kind of problem; the right rear tail lamp is out. At first observation it appears to be the same trouble, a bad light bulb. However, while making the observations on the rest of the out-

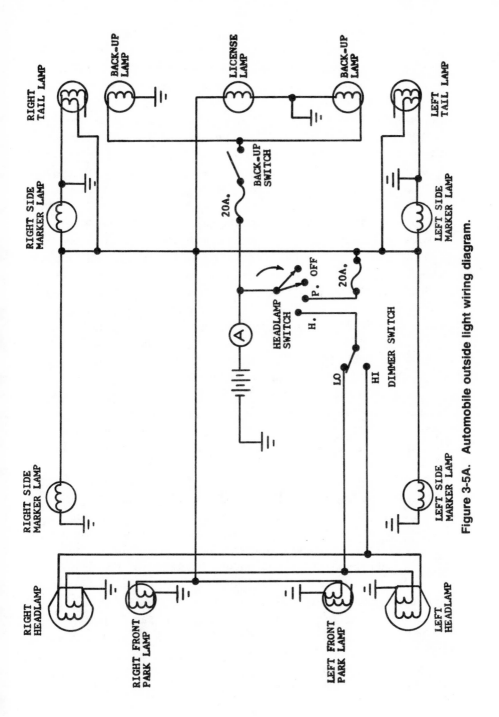

Figure 3-5A. Automobile outside light wiring diagram.

side lighting, you find that the right rear brake light does not work either. This fact changes the picture. It would be highly unusual for both filaments in the right rear tail lamp to fail at the same time. There is a common connection for them, however: the ground. Generally the tail light fixture snaps into a hole inside of the trunk compartment. The friction fit between the fixture and the car body provides the ground connection. In this case the problem could be that the tail light fixture fell out or was accidentally pulled out. Or the friction fit may have rusted or oxidized enough to destroy the ground connection. As soon as a good ground connection is restored the problem is corrected.

Suppose the right rear tail light and brake light work but only at about half brilliance. A high-resistance ground connection could cause this trouble. Or a twelve-volt light bulb in a six-volt circuit could be the problem. If the brake light is normal brillance but the tail light is half brilliance, check the voltage between the socket and ground. The VOM measures 12 VDC on the brake light but only 8 VDC on the tail light. Follow the tail light wire back and you'll find it spliced with the wire to the left rear tail lamp. However, this lamp has full brilliance. The splice must have developed a high resistance. Repairing the splice corrects the problem. Use this same kind of logical reasoning for any lamp problem of this nature in the automobile.

Now the problem is that the lighting fuse has blown. This could be a short circuit or simply a fuse failure. To find out you'll have to try another fuse or measure the current drawn with a high-current amme- ter. This fuse is generally a 20 amp fuse. Placing the ammeter leads on the fuse holder terminals with no fuse in it will measure the current. If the ammeter reads less than 20 amps with all the loads on, the problem was probably a fuse failure. For more than 20 amps you have a short circuit. Make sure the fuse clip is making good connection with the fuse. A resistive connection here could cause enough heat buildup to actually melt the fuse element from the case. It appears that the fuse is blowing, but in reality it is simply melting due to the bad connection. If you do have a short circuit in your lighting system it means that a wire, connection, or some live part of the lighting system is touching the frame. To find the short you'll have to remove the loads for that circuit one at a time until the excessive current stops. *Note:* If your car has a dashboard ammeter, this instrument will monitor the current draw instead of a test ammeter. As soon as current returns to normal, the problem is in the parts associated with the last load removed. Look for a wire whose insulation has been scraped off as it has vibrated against some sharp or ragged edge.

Sometimes car wiring problems seem more difficult than they really are because of three factors:

1. Most car wiring is bundled together in a cable or wiring harness.
2. Wires that go between the engine compartment and the passenger compartment are connected with a bulkhead connection.
3. Much of the wiring is under the dashboard making it hard to locate and test.

Keep in mind the following pointers and you can minimize these difficulties. Automobile wiring is color-coded. A blue wire with a white tracer fastened to the bulkhead connector will still be a blue wire with a white tracer when it comes out the other side. In the engine compartment the wires become dirty and greasy, covering up their colors. It only takes a minute to wipe them off to find the color of the wires you are following. If you need to check continuity of bundled wires use a VOM ohmmeter. Grounding one end of a bundled wire and measuring between the other end and the frame will verify the continuity. If you are looking for a shorted wire, make the same test but don't ground the end of the wire.

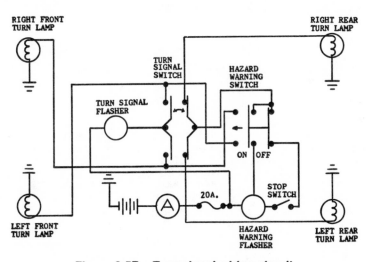

Figure 3-5B. Turn signal wiring circuit.

If you have to work under the dashboard make sure you have plenty of light. Push the front seat as far back as it will move. Try to make yourself as comfortable as possible. Removing a heater or defroster duct can give you a lot of extra room. Most instrument clusters can be removed quite easily. Disconnecting the speedometer cable, electrical plug, and removing a few screws is generally all that's necessary. Now you can work through this rather large hole in the dashboard instead of working from underneath.

Figure 3-5B shows a typical turn signal wiring circuit. As in the case of the car lighting circuits, analyze what is wrong with the turn signal circuit operation before jumping in and troubleshooting.

The following table shows some of the common problems with turn signal circuits.

PROBLEM	POSSIBLE CAUSE	REPAIR
Outside lamps work normally. No light on instrument indicator.	Burned-out indicator lamp	Replace bulb
Right or left side of system does not flash.	Bad front or rear bulb on defective side Defective contact in switch	Replace bulb Repair or replace switch
System does not flash.	Defective flasher unit Bad switch	Replace flasher Repair or replace switch
System dead.	Open fuse Open power wire Bad flasher	Replace fuse Refer to Figure 3-5B and repair Replace
Right or left outside lamps dim with no flash. Indicator lamp bright.	Poor bulb ground connection on defective side	Clean and tighten ground connector
No cancellation after turn completion.	Broken or misaligned cancelling finger in steering column Bad switch	Replace or align Replace switch

4

HOW TO DIAGNOSE AND REPAIR THE IGNITION SYSTEM FOR SMALL GASOLINE ENGINES

The electrical system in a small gasoline engine is used to produce the spark that ignites the gasoline in the combustion chamber. When the engine fails or erratic operation begins, there's an extremely good chance that something is going wrong electrically. This chapter will describe the operation of the electrical system starting with the formation of the electrical energy and ending with the firing of the spark plug.

You'll find out methods for checking the spark plug. You'll learn how to check, adjust, and replace the points and condenser. Read about the coil and find out some checks to make to see if it's okay. Solid-state ignition systems for small engines are here and becoming more popular every year. You'll find out how a capacitor-discharge ignition system works and how to repair it.

4-1. Understanding the Electrical System

Most small gasoline engines do not have a battery to furnish spark voltage. Yet you know they have a spark plug and if you have ever accidentally touched the plug while the engine was running you really know there's high voltage coming from somewhere!

Here's how it works. The flywheel, which is underneath or next to the rope starter, has a permanent magnet pressed into one side of the outside edge. Positioned very close to the circumference of the flywheel are the armature poles of a step-up coil (see Figure 4-1A). As the flywheel is turned with the rope starter, the magnet imbedded in it will move rapidly by the armature poles of the coil. The moving magnetic field will induce a voltage into the armature. The armature acting as the primary of a step-up transformer will produce a magnetic field, which cuts the many turns of wire in the coil. Since the amount of turns of wire is directly proportional to the induced voltage, a very high AC voltage can be produced. When the primary current is at maximum a cam located on the crankshaft opens the points, stopping primary current. The magnetic field collapses across the coil inducing a very high secondary voltage. A high-voltage secondary lead couples the burst of high voltage to the spark plug. A spark jumps from the center spark plug electrode to its grounded electrode. The spark ignites the combusion gases in the cylinder causing the energy to produce mechanical

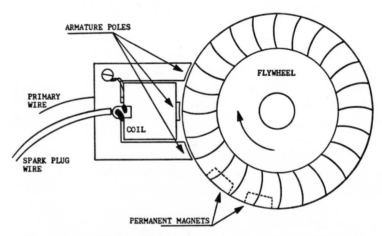

Figure 4-1A. Flywheel, permanent magnet, and coil of a small gasoline engine.

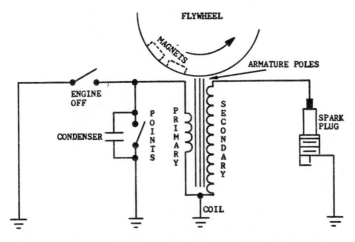

Figure 4-1B. Electrical system of a small gasoline engine.

motion. A condenser is placed in parallel with the points to prevent the points from arcing when they open and also to furnish additional energy to the primary circuit.

Figure 4-1B shows the schematic of a typical electrical system for a small gasoline engine. Generally the engine ''off'' switch is located in the carburetor controls and is activated by the throttle control cable mounted on the handle of the device.

4-2. Ways of Checking the Spark Plug

When you have trouble with a small gasoline engine, one of the first things you should check is the spark plug. Suppose your rotary lawn mower won't start. You've pulled the rope starter until your arm is tired. Check to see if the spark plug is receiving high voltage. Remove the spark plug wire from the end of the spark plug. Position the wire so that its end is about one-quarter inch from the engine (Figure 4-2A). Now pull the rope starter while watching the space between the end of the wire and the engine. You should see a fairly strong spark jump the quarter-inch gap from the wire to the engine. If you see the spark the electrical system is probably functioning okay up to the spark plug. Next, remove the spark plug from the engine with a spark plug socket wrench. Be sure you don't exert any side pressure on the plug while you are removing it because the porcelain insulator will break.

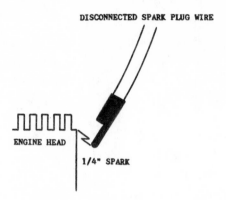

DISCONNECTED SPARK PLUG WIRE

ENGINE HEAD

1/4" SPARK

Figure 4-2A. Checking for spark with the spark plug wire disconnected.

Inspect the plug carefully. The condition of the plug can tell you a lot. Figure 4-2B shows various spark plug conditions. If the spark plug has been in operation for a long period of time, you'll see a lot of electrode wear and large deposits built up between the center core and the plug body. In this case a new spark plug is the answer.

Clean and regap the plug if it is not worn or fouled excessively. Carefully scrape the deposits away from the center and ground electrode with a small, sharp knife and wire brush. Wash with a solvent such as a carburetor cleaner or gasoline. Reset the gap with a wire-type feeler gauge usually about .025 inch (see Figure 4-2C). Tap the ground electrode on a hard surface until the gap closes to the correct distance. After the plug has been cleaned and gapped, check its operation before

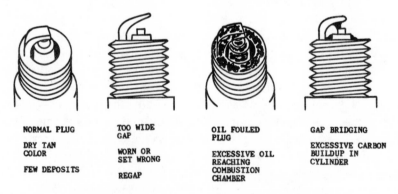

NORMAL PLUG

DRY TAN
COLOR

FEW DEPOSITS

TOO WIDE
GAP

WORN OR
SET WRONG

REGAP

OIL FOULED
PLUG

EXCESSIVE OIL
REACHING
COMBUSTION
CHAMBER

GAP BRIDGING

EXCESSIVE CARBON
BUILDUP IN
CYLINDER

Figure 4-2B. Various spark plug conditions.

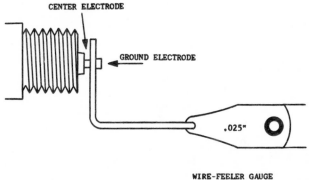

CENTER ELECTRODE

GROUND ELECTRODE

.025"

WIRE-FEELER GAUGE

Figure 4-2C. Checking the spark plug gap with a feeler gauge.

replacing. Connect the spark plug wire to the end of the plug. Ground the body of the plug to the frame while pulling the rope starter. If the electrical system is working, you'll see a strong spark jump between the center and the ground electrode. Replace the plug, tightening it just enough to compress the spark plug washer. Pull the rope starter. If the spark plug was the problem, the engine will start like magic.

Suppose when you pull the spark plug out it is gapped okay and shows no wear or deposit buildup. If it is excessively wet from gasoline, the engine firing chamber probably has been flooded. Dry the plug. Allow enough time for the gasoline to evaporate from the engine. Check the plug using the previous test. If it tests okay, replace the plug and your problems may be over.

On the other hand, if you pull the spark plug and it is absolutely dry you can be sure that gasoline is not reaching the firing chamber. If the plug tests okay, you've got carburetor problems. Very often you can help things along by shooting a blast of aerosol engine-starting liquid into the spark plug hole. Quickly replace the plug and try the engine. There's a good chance it will start to work.

It's a good idea to clean the combustion chamber in the engine especially if the spark plug has excessive amounts of lead deposits on it. Cleaning off these deposits will increase the engine power and keep the valves seating properly. It's worth the time doing this, paying off in better engine operation and longer life. Here's all you do. Remove the screws that hold the cylinder head in place (the cylinder head is that part of the engine into which the spark plug is threaded). Generally six to eight machine screws hold it in place. All the screws may not be the same length so be certain to notice where they belong. Remove the head, prying it off carefully with a screwdriver blade. Try not to damage the gasket. Pull the rope starter until the piston is at the top of the cylinder and both valves are closed. Scrape the deposits from the combustion chamber, top of piston, and around the valves; wire brush and clean with a good solvent. Pull the rope starter again to check for good seating on the valve operation. Replace the gasket and head tightening the screws evenly until they just become tight. Tighten the screws in an alternate sequence so the head has uniform pressure on it. After the engine has run for about five minutes, check and retighten the head screws in a manner that applies uniform pressure to the head.

4-3. How to Check and Replace Magneto Points and Condenser

Check the spark as previously described. If the spark is weak or nonexistent you'll have to inspect the magneto points and condenser. The toughest part of this operation is reaching them. They are most likely located underneath the flywheel. Remove the blower housing cover, three or four screws, and you'll see the flywheel. Generally the flywheel is held in place with a nut. Remove the nut. You won't be able to remove the flywheel without tools because it fits very tightly on the tapered shaft. A flywheel puller will do the job. If you don't have a puller try this technique. Screw the nut back on the shaft only until it is flush with the end. Then strike the end of the shaft and nut sharply with a soft-faced hammer, jarring loose the flywheel. For best results, read and follow the directions for flywheel removal for your particular engine from your owner's manual.

After the flywheel is removed, remove the dust cover from the points and condenser. The points should be clean and line up squarely to make good contact. Turn the shaft until the points open to the widest gap. Check the gap with a feeler gauge for proper separation. A typical gap distance is .020 inches. Generally the gap can be adjusted by moving the position of the condenser. If the points are badly pitted or burned replace them and the condenser. If they are still in fairly good shape dress them up with fine sandpaper.

You can check the condenser to some extent with a VOM. Set the VOM to the highest resistance setting. Measure the resistance from the condenser case to its terminal (Figure 4-3). The meter pointer should

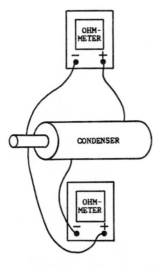

Figure 4-3.   Checking the condenser with a VOM.

show a momentary resistance, then drop back to maximum resistance (infinity). Interchange the VOM leads position and repeat the measurement. Again, a momentary resistance reading and a drop back to infinity should be observed. A good condenser will test in this fashion. A shorted or leaky condenser will indicate some resistance at all times. An open condenser will not show a momentary resistance. Even if the condenser checks okay with a VOM there's a chance that it won't work under actual operating conditions. It's a good policy to always replace the condenser when you replace the points. If you are in doubt about the condenser, replace it. You'll save yourself some time.

Before replacing the dust cover, inspect the connecting wires making sure they are still well-insulated and not shorted to the case. Many engines have a wire connected to the condenser terminal that is fastened to the engine "off" switch. In the off position this wire is grounded to the housing killing the ignition. Check this wire with a VOM ohmmeter to make sure it is only grounded in the "off" position.

4-4.   Testing the Coil

Coils are very rugged and reliable. They seldom go bad. However, occasionally either the primary wire or the secondary coil wire will vibrate enough to break or come loose. This problem is readily apparent once the cover is removed from the engine.

Another coil condition that you should notice is the gap between the flywheel and the armature poles of the coil. The armature poles should be as close to the flywheel as possible but never touch. Generally an air gap between .010 and .020 inches is recommended. An easy way to adjust the gap is with shim stock or any kind of flexible material that has the previous indicated thickness. Loosen the coil mounting screws and insert the shim stock between the armature poles and the flywheel (Figure 4-4A). Gently push the coil toward the flywheel so that both armature poles are evenly positioned against the shim stock and flywheel. Tighten up the coil mounting screws and remove the shim stock. The coil should be perfectly positioned. Spin the flywheel by hand to make certain there are no high spots on it that could rub the armature poles.

If you suspect a defective coil, you can make some ohmmeter checks that may be significant. Figure 4-4B shows the picture and schematic of a typical coil. The primary resistance is generally too low to measure with a VOM ohmmeter. It normally measures close to zero ohms. One side of the primary is usually grounded on the coil frame. The other primary wire goes directly to the ungrounded side of the

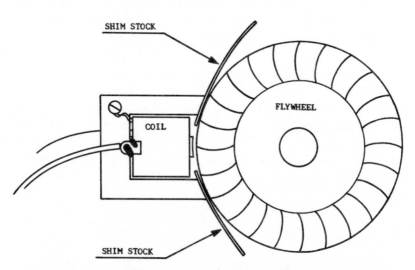

Figure 4-4A. Adjusting the air gap between the coil and fly-wheel.

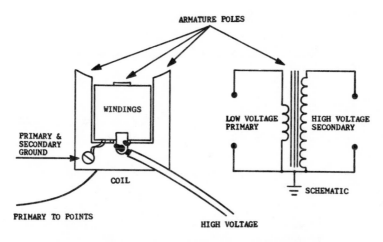

Figure 4-4B. Small engine coil and schematic.

points. A high resistance or infinity measurement across the primary indicates an open winding.

The secondary resistance can be measured from the spark plug end of the high-voltage wire to ground. You should measure a fairly high resistance depending on the type of coil, probably three to six thousand ohms. A low resistance indicates a shorted secondary; infinite or very high resistance means an open secondary. If these measurements seem okay, here's another test you can make. Most coils have one wire of the primary and secondary windings twisted together and grounded underneath a coil mounting screw. Disconnect these two wires from each other and the ground. Now measure the resistance from each of these wires to ground. Both should measure infinite resistance (if the primary measures continuity turn the flywheel to make sure the points aren't in the closed position). A resistance measurement would indicate a short to ground in one of the windings. Another check to make is between the primary and the secondary wires. Again, an infinite measurement is normal. Any resistance indicates a short between the primary and secondary.

If all of the previous resistance measurements check out okay, the coil is probably good. However, the only foolproof method to assure this is to substitute with a good coil. Make sure to first check out the rest of the ignition system thoroughly before substituting the coil. As mentioned earlier, coils are very durable and seldom become defective.

4-5. Learning About Solid-State Ignition

One of the new methods to fire the spark plug on small gasoline engines uses a capacitor-discharge ignition system. This system has no moving parts to wear out and should be more reliable than the conventional ignition.

Figure 4-5 shows the schematic of this new type of ignition system. Three coils are used. One coil is called the ''charge'' coil, another is called the ''trigger'' coil, and the third coil is very similar to the conventional ignition coil. As in the standard ignition system, a magnet is embedded into the flywheel to energize the coils. When the magnet moves past the metal armature of the ''charge'' coil, an AC voltage is induced into it. The AC is changed to DC by the rectifier and charges up the capacitor. While this action is happening the flywheel magnet is continually moving inducing another voltage into the trigger coil. The trigger voltage allows the SCR to conduct enabling the capacitor to discharge through the primary winding of the ignition coil. This pulse of energy creates a magnetic field that cuts across the secondary winding producing the high voltage, 30,000V, that fires the spark plug.

You can check out this system much like the standard ignition system. Pull out the spark plug and check it (section 4-2). If you are not sure of the plug's condition, try a new one. Position the spark plug against the engine so that the threads make a good ground. Crank the engine while watching the plug electrodes. You'll see a good spark zap across the electrodes if the ignition system is functioning. If you don't

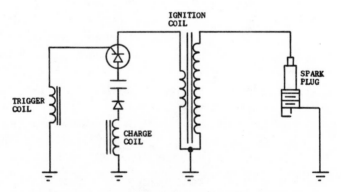

Figure 4-5.   Capacitor-discharge ignition system.

see a spark, check out the ''off'' lead that stops the engine. When this wire is grounded the ignition system is dead.

If the ''off'' wire is not grounded you may have a bad solid-state system. All the parts are built into a module that can be replaced very easily. Generally it is held in place with just a few screws. When replacing the module make sure to follow the manufacturer's recommendations as to its position in relation to the flywheel. The clearance is not the same at both ends and has to be set correctly for proper operation. "

5

SIMPLE ELECTRICAL REPAIR FOR UNIVERSAL MOTOR DEVICES
(Power Hand Tools and Small Electrical Appliances)

Most portable electric tools and small appliances use a universal motor. They are most popular on drills, saber saws, sanders, vacuum cleaners, food mixers, and sewing machines.

This chapter will discuss the construction and operation of typical universal motor devices. You will find that they are relatively simple to understand and easy to repair. Tips on checking everything electrical from the line cord to the armature are included. You probably have a few of these devices around the house just waiting for some attention.

5-1. Knowing How the Universal Motor Operates

It is called a universal motor because it can run on AC or DC. Figure 5-1A shows the electrical schematic of a typical universal motor

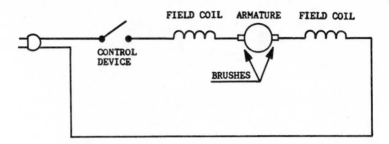

Figure 5-1A. Universal motor device schematic.

device. All the parts are connected in series with each other. As the current flows through the windings the magnetic force of the field coils will interact with the lines of force of the armature and will cause rotation. When the armature moves it drives a gear arrangement or pulley to transfer the energy to whatever the motor is driving.

Most of these devices can be disassembled quite easily. Generally they are held together with a few screws. Figure 5-1B shows a cross-sectional view of an electric drill. Notice the physical location of the field coils, armature, commutator, brushes, and bearings. It's a good idea to make a rough sketch of the parts during disassembly. You'll appreciate it when you are putting it back together.

5-2. Practical Ways to Check the Line Cord and Plug

If you use an electrical device often, sooner or later nothing is going to happen when you turn it on. The first thing to suspect is the line cord and plug. The easiest way to check the cord is to wiggle it while power is applied. First twist the cord back and forth near the plug. If nothing happens, repeat the twisting near the tool or appliance end of the cord. Generally when there's a break in the cord it will momentarily make contact running the device as long as the cord is held in a certain position. Once you identify the bad end, it's usually a simple matter to cut the bad area away and reconnect. See section 3-2 for tips on attaching plugs.

When the wiggling of the cord is not successful, you'll have to make an ohmmeter check of the cord with a VOM. A normal operating universal motor device will generally measure between twenty and fifty ohms across the disconnected plug when the switch is on. To

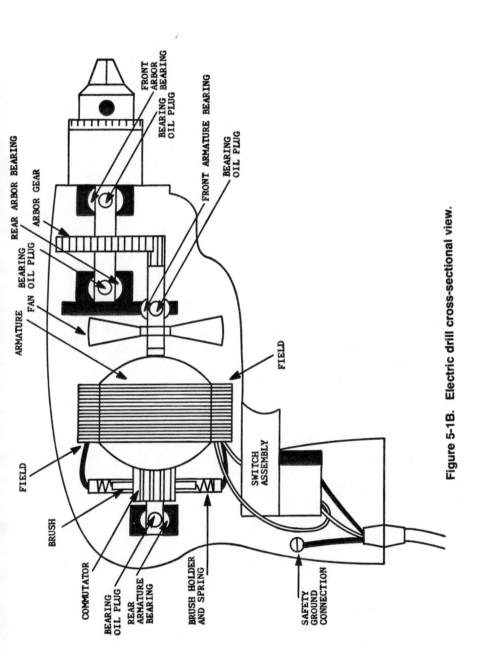

REAR ARBOR BEARING

FRONT ARBOR BEARING

BEARING OIL PLUG

ARBOR GEAR

BEARING OIL PLUG

FRONT ARMATURE BEARING

BEARING OIL PLUG

ARMATURE

FAN

FIELD

FIELD

SWITCH ASSEMBLY

BRUSH

COMMUTATOR

BEARING OIL PLUG

REAR ARMATURE BEARING

BRUSH HOLDER AND SPRING

SAFETY GROUND CONNECTION

Figure 5-1B. Electric drill cross-sectional view.

check the continuity of the line cord, fasten one ohmmeter probe to the plug and the other probe to the line cord connection in the device. If you are on the same side of the line cord the resistance should be zero ohms. Do the same for the other side of the cord. A bad line cord will measure infinite ohms on one side or the other. While checking the resistance make sure that there's infinite resistance between either side of the plug and the metal housing of the device. Any resistance measurement indicates an unsafe condition that should be repaired immediately. Some insulation in the electrical circuit has broken down permitting a path to the case. If the tool or appliance has a three-wire safety cord, there should be zero resistance between the safety ground, green wire, and the metal case.

While we're discussing electrical safety, some tools are double-insulated for safety. They usually do not have the three-wire grounded power cord. Instead, a conventional two-wire cord is used with two complete sets of insulation around every current-carrying component. When servicing this type of device be sure to reassemble exactly as it was originally or you could defeat the double-insulation feature.

If you have to replace the line cord be sure to use wire of the same gauge or heavier than the original. Wire too small could cause lack of power and overheating of the device. Make sure the cord is fastened securely to the frame of the tool or appliance with some provision made for strain relief. Choose a plug that is convenient to grip and is rugged enough to withstand a lot of use.

5-3. Testing the Switch and Control Devices

Many universal motor devices have a simple on-off switch (SPST). When switches become defective a replacement switch is usually the remedy. Occasionally a switch can be rejuvenated by spraying it with an aerosol switch contact cleaner such as that used for cleaning television tuners. These cleaners have an extension nozzle to direct the spray into the smallest cracks and crevices. Try a shot or two of spray before giving up on a switch.

If the switch is defective, it will either not turn on the device, or turn it off, or be intermittent. Very often a defective switch will not have normal mechanical action when it's activated. This is a dead giveaway for locating a switch problem. When the device will not turn on, jumper the switch with a clip lead. If the operation returns to normal the switch is bad. A switch that will not shut off the power is defective. An ohmmeter check would show zero ohms across the switch contacts for both "on" and "off" positions.

Variable-speed tools and appliances use solid-state controls. When the variable-speed feature malfunctions, you'll probably just have to replace the control unit. The majority of these devices are encapsulated in plastic or part of the internal switch or trigger mechanism. They are not made to be repaired. Just remove the leads and replace with an identical unit. Some controls have the connecting wires inserted into them with no apparent way to remove the wires. When you run into this kind of unit try inserting a 1/32-inch diameter nail or pin in the wire opening (Figure 5-3A). The pin will release the internal spring connection and you'll be able to pull out the connecting wires. To connect the wire to the new switch simply push it into the proper lead receptacle as far as it will go. As soon as the lead bottoms the connection is automatically made. You won't be able to pull the wire out again unless released with the pin.

When a variable-speed device does not work at all simply jumper the control unit with a clip lead. If the device runs full-speed, the control is probably defective. Or maybe the variable-speed appliance runs at a constant speed at all settings of the speed control. This symptom also indicates a defective speed-control unit.

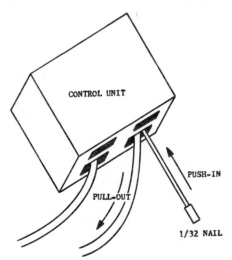

CONTROL UNIT

PUSH-IN

PULL-OUT

1/32 NAIL

Figure 5-3A. Releasing wires from a switch with a pin.

Tools that are electrically reversible generally have a switch controlling the motor direction. When the tool will work in one direction but not in the other, the reversing switch is probably the culprit. An ohmmeter check will confirm the switch condition.

On multiple-speed devices like blenders a set of push buttons are used to select motor speed. Electrically the push buttons select various taps on the motor windings. Often an encapsulated control device consisting of solid-state rectifiers is used between the push buttons and the motor to double the number of speed options. If you have a schematic handy it's easy to check out the control box using an ohmmeter. Figure 5-3B shows a typical blender-type control unit using six speeds. The rectifier-controlled position will run the motor at about half the speed of its respective direct control position. An ohmmeter connected between A and G, C and H, E and I will measure zero ohms. Placing the ohmmeter between B and G, D and H, F and I should also read low resistance. However, when the ohmmeter leads are reversed (the ohmmeter probe that was on B is put on G and the probe on G is placed on B, etc.) the ohmmeter should read a very high resistance. All rectifiers should check high resistance in one direction and low resistance in the other direction depending on the polarity of your ohmmeter leads. If the ohmmeter check confirms your suspicions of a defective control unit, replace it with an identical new unit. It won't be difficult, but just make sure you identify the leads carefully before transferring them from the defective unit to the good one.

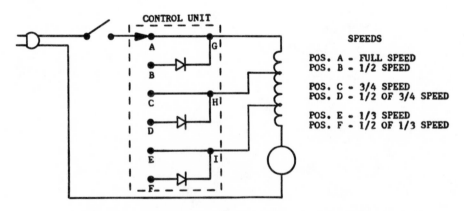

Figure 5-3B. Multiple-speed control of a blender motor.

5-4. How to Work on the Motor

The universal motor is a rugged device but needs a little maintenance once in a while. Three very common motor troubles are:

1. Lack of lubrication
2. Dirt or foreign material accumulation
3. Brush wear

Most motors use porous brass sleeve bearings surrounded by a ring of
felt. The felt is saturated with oil, which penetrates through the brass to
lubricate the motor shaft. Eventually the felt dries out, leaving no
lubrication. The shaft and bearings will become gummy and sticky
preventing the motor from turning freely. When the bearing becomes
dry enough it will likely score the shaft eventually freezing the shaft
and bearing together. Once this happens and the device is used the
windings will overheat and likely burn out.

Some devices have small holes placed in their housing so that the
bearings can be externally lubricated periodically. However, most universal motors have to be disassembled to some extent to lubricate.
Generally a few screws or bolts are all that is necessary to remove in
order to reach the felt around the bearings. Simply add a lightweight
machine oil to the felt until it is saturated. You'll be able to see the
absorption of the oil by the felt. Don't overoil since too much oil can
also cause problems. A few drops of oil can often do wonders.

Motors that operate in drills, saws, mixers, and other devices that
produce dust or shavings will eventually accumulate a lot of this residue inside of the motor. This can cause erratic operation and loss of
power. Sometimes the debris will provide an electrical path from the
live area of the motor to the metal housing causing a shock hazard. A
thorough cleaning job is all that's needed for most of these cases.

Completely disassemble the motor while making a sketch of the
parts placement. You won't believe the amount of debris that can
accumulate inside of a universal motor. An old toothbrush, tapered
toothpicks, pipe cleaners, and cotton swab sticks make great tools for
cleaning all the cracks and crevices. Be careful not to stress the connecting leads since they are usually fine-gauge wires.

The electrical connection between the rotor windings on the armature and the field coils are made with carbon brushes. The brushes are
spring-loaded so they rub on the commutator. After a lot of use the
carbon will wear down and new brushes are necessary. When the

carbon is completely worn down, the brush spring will then rub against the commutator either scoring it or shorting it out. It's a good idea to replace the brushes whenever brush wear is obvious. Most brushes are 3/8″ to 1/2″ in length when new.

At this time it's a good idea to clean up the commutator. If the motor is disassembled, one end of the armature can be placed in an electric drill (Figure 5-4A). Protect the armature shaft with a cloth or cardboard so the drill chuck will not score it. Turn on the drill while holding a strip of fine sandpaper on the commutator. The copper segments will generally clean up and shine like new in short order. Check to make sure the mica insulation between the commutator segments is below the surface of the copper. The brushes will wear out in short order if the mica is protruding above the commutator surface. To remedy this problem, simply take a fine-tooth hacksaw blade and cut the mica away until it is below the copper. If the blade is wider than the mica groove, edge-grind the hacksaw blade until it is the same thickness as the groove.

When you replace the brushes, make sure the curved area of the brush matches with the curvature of the commutator. A new brush should be fitted to the commutator by inserting a fine strip of sandpaper between the brush and commutator with the abrasive side toward the

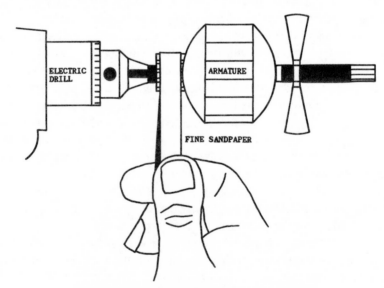

Figure 5-4A. Sandpapering a commutator.

brush. Turn the armature by hand until the brush curvature matches the commutator.

Some ohmmeter checks you can make in the universal motor are shown in Figure 5-4B. Ohmmeter #1 and #2 will measure each field coil resistance. You'll measure a low resistance here probably less than ten ohms per coil. Ohmmeter #3 checks the continuity of the brushes and the armature. Again, a low resistance is normal, ten to twenty ohms. Move the armature by hand while watching ohmmeter #3. You should see the resistance change slightly as the commutator segments move by the brushes. Ohmmeter #4 measures the series resistance string including the two field coils, the two brushes, and the armature winding. This measurement should be the same as the addition of ohmmeter readings #1, #2, and #3. Ohmmeter #5 should read infinite resistance. There should be absolutely no continuity between the metal frame or case and the electrical circuit.

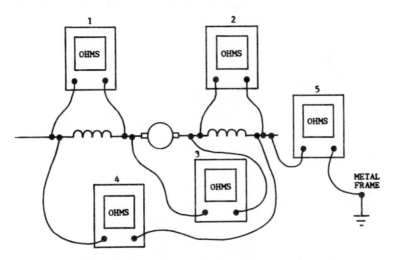

Figure 5-4B. Universal motor ohmmeter checks.

Figure 5-4C shows how to check the resistance of the commutator segments and rotor coils. Place one ohmmeter probe on a segment and the other probe on the segment directly across from it. You will measure a low resistance. Make this same check with each set of segments. Each pair should read the same resistance. No segment should have any continuity to the metal shaft.

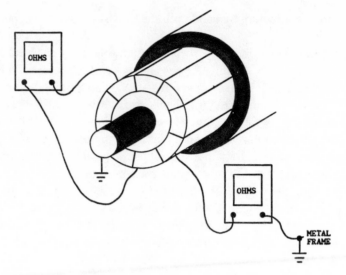

Figure 5-4C. Checking commutator segments with an ohm-meter.

6

MAKING
ELECTRIC DISHWASHER REPAIRS

When the dishwasher breaks down, you can bet that the problem is electrical in nature. Once you have a basic understanding of dishwasher operation, the repair is usually quite easy.

The electrical timer is the operating brain. It determines the sequence and timing for all dishwashing functions. Read about the timer and how to troubleshoot it in this chapter. You'll learn about solenoid valves and how they automatically control the water for the various cycles. The testing and repairing of the electric pump is described here, as well as how to diagnose the malfunction of the main motor.

All electric dishwashers become defective sooner or later. When they do, it will cost a lot of money to have them repaired. You'll find by reading this chapter that dishwashers are not difficult to understand and you can fix them a lot easier than you might have imagined.

6-1. Tips in Dishwasher Operation

Load the dishes in, fill the soap dispenser, set the timer and close the dishwasher door; we're ready for action. From now on the timer mechanism will take charge of all the valves, pumps, heaters, motors, and other items necessary to clean dishes. The timer is simply a set of switches automatically opened and closed at precise intervals. Once the timer is set at start, closing and locking the door will act as the power-on switch.

All dishwashers do not operate exactly the same. However, once you understand the operation of a typical machine, it will be easier to repair all types. Generally the first few minutes are allocated by the timer for pumping operation. The electric pump receives power through the timer contacts so it may pump any accumulated liquid out of the machine. After the pump interval, an electric solenoid water valve is actuated so hot water may enter the dishwasher. This interval is precisely timed also so just the correct amount of water will be allowed in. While the water is coming in, the main motor is turned on, spinning the impeller around so water is hurled throughout the inside of the dishwasher. Generally at this time an electric heater is also turned on heating the water to a higher temperature than is available from the hot-water tank. If the washer has a pilot light display, the first rinse light will be on.

After a few minutes of operation the action is stopped long enough for the pump to remove all of the dirty rinse water. Then new water is allowed in and soap is automatically dispensed. The soapy water is again heated and swirled around the dishes. The pilot light display will show "first wash." After a set interval, the washing will stop and the water will be pumped out again. As soon as the pump-out cycle is completed, new water will flow in and the second rinse action will begin. This will be indicated by a "second rinse" light. After the second rinse time is over, the rinsing action will stop and the soiled water will be pumped out.

New water is allowed in, the second application of soap is dispensed, and the second washing begins. The second wash lamp turns on. After about five minutes of washing, the cycle repeats for usually a third and fourth rinse interval. Generally on the fourth or last rinse the timer momentarily stops until the rinse water is heated to maximum temperature. When the last rinse interval is finished all the water is again pumped out. No new water is allowed in since this is the drying cycle. The heater and motor impellers are switched on, heating and moving the air around the dishes for about twenty minutes. Next, the pump is activated to remove any residual water and the timer turns off.

Most dishwasher cycles take forty to fifty minutes to complete from start to finish. They will usually use around twelve to fourteen gallons of hot water and need about twelve hundred watts of electricity. The previous explanation describes the basic operation of all electric dishwashers. Naturally some models are more expensive and have a few more "wrinkles" than others but the basic cleaning operation is similar.

6-2.   Understanding the Timer

All kinds of strange symptoms can show up if the timer breaks. Since the timer controls the sequence and operation of every major dishwashing action, it's apparent that a malfunction of any timer switch can create a lot of confusion. If you suspect timer trouble, your best bet is to consult the timing diagram for that particular dishwasher. Figure 6-2A shows a typical timing diagram. Notice how precise times are allocated for each of the functions. Here's an example of how you would read the diagram for the first part of the cycle. After starting the timer there's a two-minute adjustment period. During this period the first rinse light, the main power, the drain pump, and the timer motor are all on. Next, the first rinse cycle begins. During the first minute of this cycle the water valve will be activated as well as the impeller motor and heater. The drain pump will turn off. Everything else that was on will stay on. After the first minute of this cycle the water valve will close; everything else will stay the same for another minute. Then the last phase of this cycle will begin (drain). The pump will come on for a minute to remove the water. Everything else remains the same. Next, the first wash cycle starts. The water valve, first wash light, main power, impeller motor and heater, and timer motor are on; the first rinse light goes off as well as the drain pump. Follow this same type of logic for the remainder of the timing diagram.

To check the timer operation, monitor the timer sequence shown on the timing diagram with a watch as you run through a cycle. For instance, suppose the pump is not removing the water properly. You don't know if the trouble is in the timer or the pump. Connect the VOM set for 110 AC volts across the pump motor's connections. Start the timer and time the pump motor operation. At the proper interval the VOM should indicate whether or not the motor is receiving voltage. If it is getting power, the voltage should remain as long as pro-

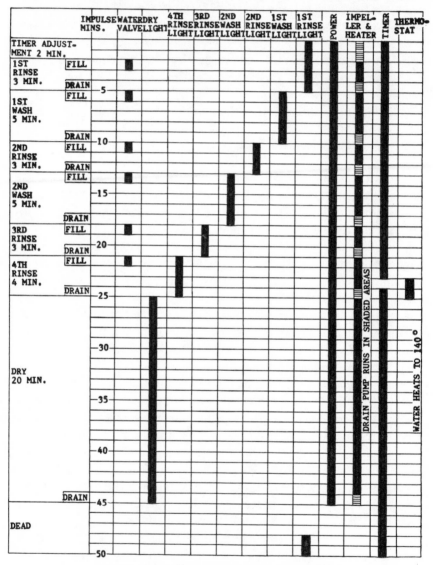

Figure 6-2A. Dishwasher timing diagram.

grammed for in the timing diagram. If voltage is present for the correct interval and the pump is not removing water, the problem is in the pump or associated drain lines. However, if no voltage is present, the problem is in the timer switches. This type of voltage measurement can be used to isolate most timer-related troubles.

Figure 6-2B shows the switch contact arrangement for a typical timer. Each switch contact is activated by an internal mechanical action as the timer motor rotates. To check continuity of the contacts, measure them with an ohmmeter as you move the timer manually through its cycle. Good contacts will measure zero resistance when they are closed.

A common timer problem is the timer motor. The motor winding can open or a connection can break causing no timer rotation. If the timer is not rotating check the AC voltage on its motor. It should measure 110 VAC. If the voltage is present, you've probably got a bad timer motor. Confirm your suspicions by checking the resistance across the timer motor terminals. Motor resistance should measure approximately one thousand ohms. In most timers the motor can be replaced without replacing the entire timer assembly.

If you have to replace the timer, make certain that you know where the various leads are fastened. The leads are generally color-coded and terminals identified in some kind of fashion. It's a good idea to make a sketch of the connections before any leads are taken off. The leads are usually quite easy to remove since they are fastened with push-on connectors.

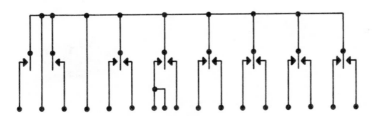

Figure 6-2B. Timer switching contacts.

6-3. Working with Solenoid Water Valves

Most dishwashers have just one solenoid water valve that controls water input to the machine. Basically a solenoid is a device that changes electrical energy to mechanical motion. It consists of a coil of wire supported by a steel frame. Inside of the coil is a movable metal plunger. When electricity flows through the coil, a magnetic force is produced in its center. The plunger is attracted to the center of the coil rapidly with a good deal of force by the magnetism. After the current stops flowing, the plunger is generally pulled out of the coil by a spring or by gravity.

In the dishwasher the solenoid is connected to a water valve in the intake water line. When the timer signals, current will flow through the solenoid coil energizing the plunger valve and permitting water to enter. After a preset interval the timer stops solenoid current, returning the plunger back to its original position and shutting off the water supply. (Figure 6-3A)

It's easy to check a solenoid water valve. Any time water is not entering the machine the solenoid is a prime culprit. Place the VOM, set for 110 VAC, across the solenoid's terminals. Start the timer while monitoring the VOM and the timing diagram. If voltage is switched to the solenoid at the correct interval but you don't hear valve action or no water enters the machine, the solenoid is probably defective. Remove

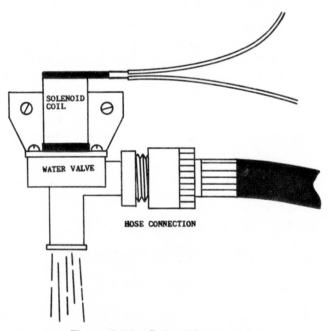

Figure 6-3A. Solenoid water valve.

power and measure the resistance of the solenoid coil. The resistance measurement will indicate the condition of the coil.

Solenoid Coil Resistance	Probable Cause	
Zero ohms or very low	Shorted coil	—replace
Infinite or very high	Open coil	—replace
Approximately 400 ohms	Normal	

As a last resort if you are still in doubt as to whether the solenoid water valve is good or bad, try this. Pull off the leads to the coil. Connect the coil directly to the voltage source it is rated, usually 110 VAC (see Figure 6-3B). Be sure the line cord clips are not touching anything but the solenoid coil terminals.

Now if you still don't get solenoid action there is no doubt that it is defective. Remove the hose or hoses, unscrew a couple of fasteners, replace the solenoid water valve with a new unit, and your problem is over.

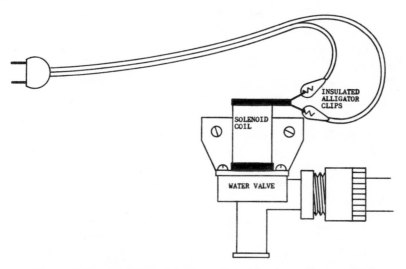

Figure 6-3B. Operating a water solenoid valve with an external power source.

6-4. How to Service the Electric Pump

The pump in most dishwashers consists of a small electric motor that drives an impeller to pump water out. The pump assembly is located at the beginning of the drain line. The motor is electrically activated by a timer switch.

There are generally three kinds of pump troubles:

1. Water leakage
2. Weak pumping
3. No pumping

The water leakage problem is not electrical in nature, yet water can be sprayed into electrical areas causing secondary problems. Tightening the pump housing screws or replacing the gaskets will usually stop most leaks. If the pump housing is made of plastic it will occasionally crack or warp requiring a replacement housing. The hose connections sometimes leak. Cutting an inch or two off the defective hose end and reclamping will generally stop this kind of leak.

A pump that is not pumping the required volume of water out of the machine will upset the entire washing and drying cycle. Soiled water will be left in the machine after every drain cycle. The dishes will not be cleaned or rinsed properly. Also, since water will be left in the machine during the dry cycle, the dishes will never dry.

Weak pumping action is most often caused by a nonelectrical defect such as a semiblocked intake line to the pump. Occasionally foreign material will lodge in such a fashion as to impede the pump impeller. After many years of use the rubber drain hoses sometimes swell internally causing an apparent pump problem. The pump motor could become sluggish due to poor lubrication. Turn the motor shaft manually. It should spin with little resistance. If it turns hard try oiling the motor bearings. When oiling is not successful, you'll have to completely disassemble the motor and pump to locate the reason for hard turning.

Another possible cause of weak pumping action is low voltage to the motor. Defective timer contacts or wire connections could result in low voltage on the pump motor. Measure the voltage with a VOM. It usually is 110 VAC. If you do measure low voltage, check the voltage across each wire connection and timer switch associated with the pump (Figure 6-4). A good connection or a good switch contact will drop no voltage. Any connection or switch that is dropping voltage will be hot to the touch. Most defective connections and splices can be repaired by a thorough scraping, cleaning, and tightening. A defective timer switch will necessitate a new timer assembly.

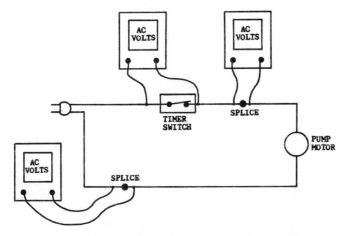

Figure 6-4. Checking voltage drop across connections and switches.

A no-pumping symptom is the easiest type of trouble to find. Either the voltage is not getting to the motor or the motor is not turning the pump. If the voltage is not present at the motor, trace back through the circuit with a VOM until the defect is located. When the proper voltage is measured on the pump motor and the motor is not turning, you've got a bad motor winding. Remove the leads and check the motor resistance. A good motor will usually measure 10 to 25 ohms. The resistance reading will confirm whether the pump motor is good or bad.

6-5. Testing and Replacing the Motor

The main motor on the dishwasher is generally a 1/3 or 1/2 horsepower 115V motor. It drives the impeller that moves the water and air for all of the cleaning, rinsing, and drying operations. Capacitor and split-phase type motors are used. A capacitor motor can usually be identified by a metal shield housing the capacitor on the side of the motor. For the rest of this discussion we will assume the motor is a one-speed motor. Testing and replacing of multiple-speed motors is covered under automatic washing machines, section 7-5.

The motor has two windings, start and run, in parallel with each other (Figure 6-5). The start winding is switched by a centrifugal

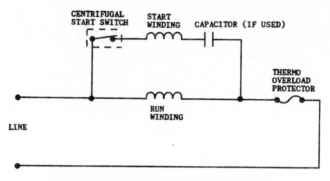

Figure 6-5. One-speed motor schematic.

switch inside of the motor housing. To enable the motor to start, current flows through both windings producing enough magnetic action for the moving part of the motor (rotor) to begin spinning. After a couple of seconds the start winding is no longer needed for motor operation. It is switched off by the centrifugal switch when the rotor reaches about 1,100 RPM. The rotor continues running from the magnetic field produced for only the run winding.

Most motors are equipped with automatic reset thermo overload protectors. This protector may be built in or mounted outside of the housing. The built-in protector cannot be replaced. You can replace the external unit. The protectors are designed to heat up and switch off when a motor overload occurs. They automatically reset after the heat returns to normal.

Motor failures are often readily apparent. The defective motor will generally cause the house fuse or circuit breaker to shut down dishwasher power. Often the windings will burn causing a lot of smoke and odor before power shutdown. It's an easy trouble to spot but an expensive one for the pocketbook. Check the mechanical operation of the impeller. It should spin freely and easily. Check for water leaks between the motor seal, gaskets, and the tub. A small amount of water leakage can cause a large amount of trouble. If the impeller spins freely and you can't find any leaks, the chances are the motor became defective without any external causes. Once a motor starts smoking it is usually defective beyond repair. The varnish insulation on the motor windings has overheated and deteriorated.

Another common motor problem results from a defective centrifugal switch or starting capacitor. In this case the motor will not start; the impeller will not move. You'll hear a buzzing noise in the motor and its housing will get hot to the touch. The centrifugal switch is not making contact or an open capacitor is causing no current flow in the start winding. Current is flowing through the run winding causing the overheating but is not strong enough to get the rotor moving.

This type of motor trouble can be repaired if you want to tackle the job. If it's a capacitor motor replace the capacitor. If that's not the problem the motor has to be disassembled. Most motors are held together with four long machine screws and nuts. Removing the nuts will enable you to remove the end section of the housing to reach the switch. Once you find the switch you'll see that the contacts have been burned away. The trick now is to find a replacement centrifugal switch. Many dishwasher manufacturers do not make their own motors. Consequently, they don't stock any internal motor parts. They will only sell you the entire motor. However, you should be able to obtain the switch from the actual motor manufacturer (stamped on the housing) or a local motor rebuilding organization. If you do replace the switch, be sure the motor runs freely manually after it has been reassembled. Also be sure to relubricate the bearings. The final check is to apply 110 VAC on the motor terminals. The motor should start up and run smoothly. You should have a big grin on your face!

If you decide to replace the motor, be sure you keep a record of the order of parts disassembly and the lead connections. Generally there are quite a few gaskets and seals to remove before the motor can be pulled out. Even some of the screws may have individual rubber seals under their heads. Don't mix these up with screws with no seals. It's a good idea to replace the gaskets and seals when you replace the motor. It's cheap insurance for a water leak. Usually the replacement motor will have detailed installation instructions packed with it. Be sure to read and follow them thoroughly since some dishwashers have critical distances between the mating of parts. After the motor is replaced, spin the impeller manually for easy operation with no binding. All that's left is testing for dishwasher operation.

7

HOW TO FIX
ELECTRICAL TROUBLES ON
AUTOMATIC WASHING MACHINES

The operation of the automatic washing machine may seem complex and difficult to understand. In reality this is not the case. There are only a handful of electrical parts that are the key to the whole operation. In this chapter you'll learn about how these parts function and how they work together. The next time your washing machine breaks down you won't have to pay the repairman a lot of money.

An understanding of automatic washing machine operation will enable you to locate and repair most electrical failures. You will learn about the correlation between the cycle of operation and the electrical circuits and become familiar with the timer and the timing chart. You'll find out how the solenoids electrically control many mechanical operations. Did you ever wonder how a washing machine knows when there is enough water in it? Here you will read about and learn how to fix water-level problems. Most washing machines operate with multiple-speed electric motors. If you learn about these motors, you can turn the difficult into the easy.

7-1. Guide to Automatic Washing Machine Operation

Here's how a normal wash cycle operates from start to finish. The user will set the wash time, water temperature, water level, and add the detergent. As soon as the timer is started the machine will be fully automatic until the clothes are cleaned, rinsed, and spun dry. Then the machine will turn itself off.

As in the electric dishwasher operation, section 6-1, the timer is the brains of the entire operating cycle. When the timer is started, it immediately turns on the solenoid valves, which permit water to enter. The water temperature is controlled by regulating the mix of hot and cold water. After the tub has filled to the desired level, the water-level switch will shut off the water solenoid valves.

The water-level switch will also turn on the main motor and agitator. The wash cycle begins and continues for the prescribed amount of time. Then the timer shuts off the agitator and turns on the pump, which removes the soiled water from the machine.

When all the wash water has been pumped out, the spin-rinse cycle begins. The water-level switch has sensed the absence of water and allows the timer to start the spin cycle. Usually during the spin cycle a number of short spray rinses enter the machine. The timer turns the water solenoid valves on and off for these spray periods.

Next comes the deep rinse portion of the cycle. The timer stops the spinning and opens the solenoid water valves. As soon as the tub is filled, sensed by the water-level control, the agitator will be turned on again. After the period of agitation, the power is transferred from the agitator to the pump. Rinse water is pumped out.

The final spin cycle begins. The same action occurs as in the previous spin cycle. When this cycle is completed, the timer turns off power to all circuits. The only thing left is to remove the clean clothes from the machine.

7-2. Working with the Timer

The best way to understand how the timer works is by referring to the timing chart showing the timer's switching operation (Figure 7-2A). The number of the timer switches are shown on the top horizontal column. The next lower horizontal column shows the functions of each switch. The next lower column indicates the terminal designation.

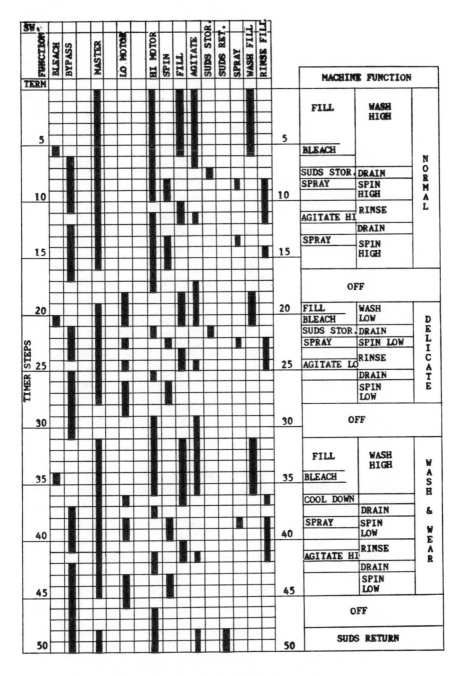

Figure 7-2A. Timer switch sequence chart.

AUTOMATIC WASHING MACHINES / 97

The fifty horizontal columns represent the timer steps. Each step is equal to two minutes duration. Let's analyze the first step of the timer operation while referring to Figure 7-2A. The black vertical bars in the switch columns show which switches are activated. As soon as the timer is started the master switch is closed. The high motor switch is closed as well as the fill switch. Also closed are the agitate and wash-fill switches. The timer switches will stay the same for five steps. On the sixth step the bleach switch is turned on for two minutes. The seventh step turns on the bypass switch; turns off the fill and wash-fill switches. The rest of the timer switches that were on stay on. The remaining steps operate in a similar fashion.

Referring to the portion of the chart labelled "Machine Function" will indicate the sequences for the three cycles of operation and the suds return feature. Notice the timer is programmed to stop after completing either a normal, delicate, or wash-and-wear cycle. Looking at the "normal" part of the "Machine Function" you can see that five steps are allocated for fill time, one step for bleach. At the same time the machine is set for wash with the agitator on high speed. After seven steps wash time is over. Step eight drains the wash water or stores it externally if the machine is equipped with a suds return feature.

The next two steps are spin periods with one step of spraying. The main motor is on high speed for spinning. Deep rinse and agitation begins on step eleven. Next is the drain cycle and finally the last spin-dry with one spray step. The motor is again on high-speed operation. Then the machine shuts off on step seventeen.

Let's take a typical timer step and correlate it with the electrical schematic of the washing machine. Refer to Figure 7-2A while looking at the schematic, Figure 7-2B. Here's what happens electrically, shown with arrows, as the first timer step is operating. The manual on-off switch activates the master timer switch and completes the circuit to the console lamp. Electricity then travels through the "off-balance" switch (the switch will open automatically if the basket spins unevenly) to the water-level switch, which is in the empty position. The next switch path is to the timer "fill" switch and through the timer "wash-fill" switch. The water temperature switches are set manually for the desired water combinations, hot wash and cold rinse. The current will find a path through the hot water valve solenoid and back to the other side of the line. Notice that timer switches for the high motor and the agitate solenoid are on but no current can flow through

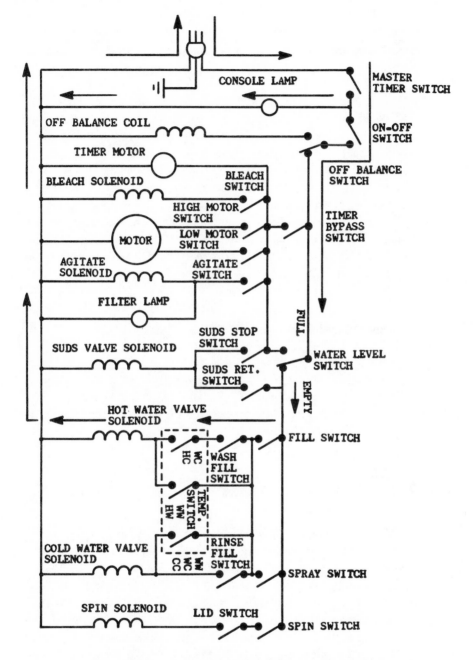

Figure 7-2B. Automatic washing machine schematic.

them until the water-level switch senses that the tub is full. The first timer step, although appearing complex, only activates two loads, the console lamp and the hot water solenoid. Any timer step can be analyzed in this fashion using the timing chart and the electrical schematic.

If you suspect a timer switch failure, it's easy to troubleshoot. Set the VOM for 110 VAC. Connect one VOM lead to the common line (white) in the machine. Touch the other VOM lead to the wire coming from the suspected bad switch. Measuring 110 VAC will indicate that the switch is closing and is probably okay. Measuring 0 VAC will indicate a bad timer switch. Make certain that the particular switch you are measuring is programmed to be closed.

To check a bad timer motor simply measure the voltage across its terminals. If 110 VAC is present and the timer is not moving, the answer is a bad timer motor. You probably won't have to replace the entire timer assembly since most timer motors can be replaced separately.

7-3. Testing and Replacing Solenoids

Solenoids are used throughout a washing machine since they are efficient devices to change electrical power into mechanical motion. All washing machines use solenoids in varying quantities depending on the type and complexity of the model (see Figure 7-3). Here are some examples of devices controlled by solenoids:

1. Water-control valves
2. Agitator
3. Spin drum
4. Two-way valve
5. Dispensers

All solenoids depend on the principle of magnetism attracting a movable metal rod or shaft (see section 6-3). When a solenoid is working correctly you can hear it and in most cases see the action. This fact makes it easy to spot a bad solenoid-operated device. For instance, suppose water is not entering the machine during the fill cycle. If the external water supply is turned on (you wouldn't believe how many water troubles result from turning off the outside faucets), there's only a few things that could cause this problem: a timer switch, the water-

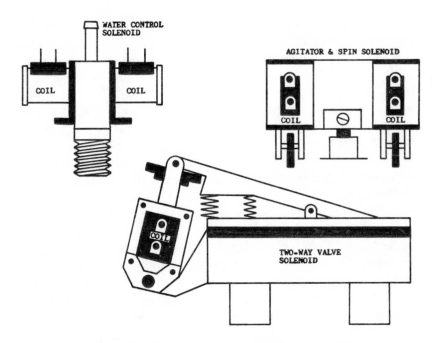

WATER CONTROL
SOLENOID

COIL   COIL

AGITATOR & SPIN SOLENOID

COIL   COIL

COIL

TWO-WAY VALVE
SOLENOID

Figure 7-3. Typical washing machine solenoids.

level switch, bad connections or wire, and the water solenoid valves. If there's no solenoid action, measure the voltage across the solenoid terminals with a VOM. If you measure 110 VAC, it looks like a bad solenoid. If you measure 0 VAC the trouble will be a switch, wire, or connection.

When the voltage check indicates a defective solenoid, you can verify the conclusion with one or two other checks. One, make a resistance measurement across the solenoid terminals with one of the leads removed. A good water valve solenoid coil will probably measure around four or five hundred ohms. Two, pull both connecting wires off the solenoid coil and power the coil with a separate 110 VAC source. If the solenoid still doesn't work, you can be absolutely certain that it is bad.

All solenoid coils do not measure the same resistance. Use this rule of thumb for evaluating resistance measurements: The larger the coil, the smaller the resistance and vice-versa. As previously indicated, the water solenoid coil would measure about four hundred ohms. This is a small coil not doing a great deal of work. However, a solenoid coil that is connected to a lot of apparatus is going to need a lot of current to operate and its resistance will be low, it could be as low as ten ohms.

A lot of washing machines have two solenoids that operate in conjunction with cam bars, mechanical devices that determine if the machine is agitating or spinning. Check these solenoids with the VOM just as you would the water valve solenoid. Machines that have a suds return system use a two-way valve controlled by a large solenoid. This enables the timer to have electrical control on the drain system. In one valve position the water will drain and in the other position the suds water will be channeled to the external reservoir. If you're having drain problems, this solenoid is a prime suspect to check.

Solenoids are used to automatically dispense fluids in rinse and bleach dispensers. The solenoid plunger is fastened to the drain stopper in the dispenser reservoir. During the proper interval in the timer cycle, the solenoid energizes "pulling the plug" for the dispenser. To check the action manually operate the solenoid while inspecting operation. Often the only fix needed is a good cleaning of the reservoir and plug mechanism. When everything seems to be correct, mechanically okay, check the solenoid coil voltage during the proper timer interval. Replace the solenoid when the voltage is present but there is no solenoid action.

7-4. Ways to Check Water-Level Controls

The water-level control determines the water level in the tub. As soon as the proper water level is reached, a circuit is completed that allows washer action. Most water-level controls are switches activated by air pressure. Figure 7-4A shows a typical water-level pressure switch installation.

When the tub fills with water, air is trapped in the tube leading to the pressure switch. As the water rises in the tub the air pressure increases in the tube. With sufficient pressure a diaphragm in the switch activates the contacts shutting off water to the machine and allowing the agitator to start. After the water is pumped out of the machine the air pressure decreases and the diaphragm and contacts return to their original position.

Generally there are three different types of water-level controls: single-level, multi-level, and infinite-level. They all operate on the same air pressure principle. A mechanical arrangement changes the

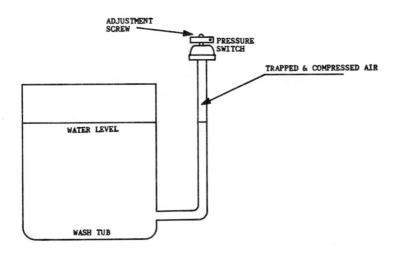

Figure 7-4A. Water-level pressure switch.

pressure needed to activate the diaphragm on the multiple and infinite type switches. Some switches are equipped with an adjustment screw if they don't activate at the proper water level. A very slight adjustment is usually all that is needed.

Figure 7-4B shows the schematic of a water-level switch and how to check it with a VOM. To troubleshoot the water valve circuit, measure the voltage between point X and the white wire. Measuring 110 VAC here indicates master timer, on-off, and off-balance switches are okay. Move the VOM probe from point X to Y. You should measure 110 VAC here until the tub is filled. Voltage at X but not Y indicates a defective water-level switch. When the tub is filled the voltage should move from Y to Z.

If voltage measurements indicate a bad switch, double check the switch with ohmmeter readings. Pull off all the switch wires. The resistance should be zero ohms between points X and Y; infinite ohms between X and Z. Remove one end of the air pressure hose and blow into the switch. The ohmmeter should now read zero ohms between points X and Z; infinite between X and Y. Replace the switch if it doesn't pass the ohmmeter test. In the event the voltage check indicates a bad switch but the resistance measurement is okay, the problem is probably a blocked air pressure line.

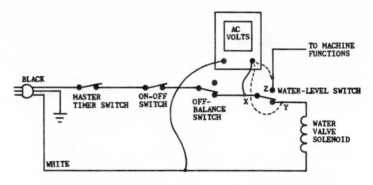

Figure 7-4B. Checking a water-level switch with a VOM.

7-5. Checking the Motor

Machines with only one operating cycle use a single-speed motor. Checking this kind of motor is covered in section 6-5. Many washing machines have two or three operating cycles requiring multiple-speed motors. This section will discuss the operation and testing of these motors.

The split-phase and capacitor motors are common types used in washers. Figure 7-5A shows the schematic of a typical two-speed motor. These motors have three windings: start, normal, and slow. In the normal position the timer applies voltage to the start and normal windings. Two or three seconds after the motor starts, the centrifugal switch inside the motor activates opening the start circuit. The motor will continue to run on only the normal winding.

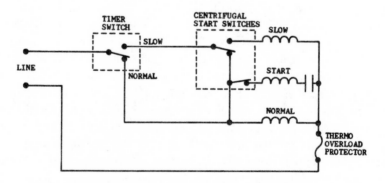

Figure 7-5A. Two-speed motor schematic.

104 / AUTOMATIC WASHING MACHINES

When the user selects the slow speed, power is connected to the slow-speed timer switch to the start and run windings. The motor will start as before. However, now when the centrifugal switch activates, the power will be transferred from the normal winding to the slow winding; the start winding will open as usual.

Basically the same sort of action occurs in the three-speed motor shown in Figure 7-5B. An extra slow winding is added plus another set of contacts in the centrifugal switch. All the windings operate exactly the same as for two-speed operation. For extra-slow speed the timer directs power to the extra-slow motor terminal. The motor starts as usual with the start and normal windings. The centrifugal switch then activates transferring the power to the extra-slow winding and disconnects the others.

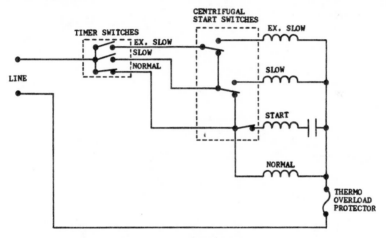

Figure 7-5B. Three-speed motor schematic.

When you suspect motor trouble measure 110 VAC at the motor terminals; one lead of the VOM to the common terminal and the other to the start and run terminals. If you measure the voltage and the motor is not running, it is defective. A bad centrifugal switch or capacitor would prevent the motor from starting; see section 6-5 for details.

There are many types of motor failures most of which are difficult to repair without special training and equipment. There are motor rebuilding shops but generally the price between a new replacement motor and rebuilding the old one is very close. Here are some of the things you can do to repair a motor.

1. Repair or replace the centrifugal switch.
2. Replace the capacitor.
3. Lubricate the bearings.
4. Repair or replace the thermostat overload protection.

Refer to Figure 7-5C. An ohmmeter check can confirm an open circuit or thermostat in a split-phase or capacitor-type motor. Set the VOM on R × 1 range and measure the resistance between motor terminals 1 and 2. A low resistance measurement, 1 or 2 ohms, indicates that the "normal" winding and the thermostat overload protection have continuity. A high resistance reading would indicate bad contacts in the thermostat or an open "normal" winding, or poor connections. Measure the resistance between terminal 1 and the motor case. There should be no resistance (infinite). A resistance reading would indicate that some part of the circuit is shorting to the motor case.

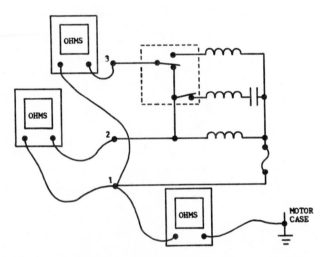

Figure 7-5C. Ohmmeter measurements on a two-speed motor.

Now measure the resistance between terminals 1 and 3. Again a very low resistance measurement is normal. A high resistance reading would probably indicate bad centrifugal switch contacts as long as the previous measurements were okay. If all the resistance measurements seem to be okay but the motor does not start, the problem is in the starting winding circuit. It's tough to check the continuity of the starting winding since it's in parallel with the run winding. The ohmmeter is not sensitive enough to detect the difference between the single winding or two in parallel. In a capacitor start motor the capacitor will block the ohmmeter path through the start winding.

When the motor hums but does not start, something is probably wrong in the start circuit. First make sure the motor turns freely by hand. If it's mechanically okay then you can check the centrifugal switch action by spinning the motor by hand while power is applied. If the motor continues to run after it is hand started, the centrifugal switch needs attention. A defective capacitor will also keep the motor from starting. Try a replacement capacitor.

8

FIXING THE AUTOMATIC
ELECTRIC CLOTHES DRYER

In this chapter you will read about the basic theory of operation for all automatic electric clothes dryers. You'll find out how a few key electrical items can operate and control all dryer functions. As in most major automatic electrical appliances, the timer directs all operations. You will become familiar with a typical dryer timer.

When the heating system fails, a few checks are all that's necessary to locate the problem. Heating unit defects can be diagnosed with ease. You'll learn how the dryer knows the required temperature and how you can troubleshoot these heat-sensing devices. All dryers need a blower to circulate the heated air through the clothes. This chapter will acquaint you with simple methods to remedy blower troubles. Dryer motors can go bad and often do. You'll find out things to check that are associated with malfunctioning motors.

8-1. Theory of Operation

Here we go! First throw in the wet clothes and shut the dryer door. The closed door will set an interlock switch allowing the dryer to start when commanded. Next, the dry temperature control should be set for the recommended setting of the fabric being dried. This control will select thermostatic switches inside of the dryer that will maintain the required temperature by turning on and shutting off the heat supply. Set the timer control for the amount of time needed for the fabric being dried. Switches inside of the timer will control operation during the drying cycle. All that's left is to push the start switch.

The timer will start the drive motor, which turns the blower and the drum. At the same time another set of timer contacts will activate the heating mechanism. The heating unit is part of a series circuit consisting of the timer switch, an operating thermostatic switch, a safety thermostatic switch, and centrifugal switch contacts in the motor (see Figure 8-1). All these contacts must be closed in order for the heating system to function.

The operating thermostat will close and open depending on the setting of the console dry temperature control. The safety thermostat is necessary to shut down the heating mechanism if the exhaust outlet becomes clogged with lint or other overheating occurs. The centrifugal switch contacts are on the motor but are not part of the motor's electrical circuit. These contacts are necessary to insure that the motor is running before the heat is applied. As soon as the motor reaches about three-quarters running speed, centrifugal force will close the contacts allowing a current path for the heating circuit. Any time during the

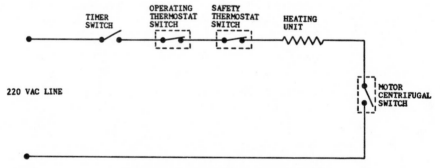

Figure 8-1. Heating unit series circuit.

operating cycle, opening the door will break the circuit to the motor and shut off all dryer operation. The door switch is in series with the drive motor. Also, as soon as the drive motor stops, its centrifugal switch will open the heating circuit.

Near the end of the timing cycle the timer will shut off the heating circuit but allow the blower and drum to continue for five or ten minutes using unheated air for cooldown. This will allow the clothes to be cool enough to touch and is necessary for permanent press fabrics. Some machines have a buzzer that sounds at the end of the drying operation. This buzzer is generally connected through the centrifugal switch so that it is automatically disconnected when the motor coasts down.

The buzzer will generally be audible for a few seconds. Now all you have to do is open the door and remove the dried clothes. When everything is working okay this is the typical operation of automatic dryers. Once you understand the theory of operation, you won't have any trouble locating the defective area when a malfunction occurs.

8-2. Troubleshooting the Timer

The timer generally consists of two basic components: the motor-gear assembly and the switch box. The small 110 VAC motor is geared down to turn a cam arrangement, which activates switches during periods of its rotation. There are usually three kinds of timers in use: single-cycle, dual-cycle, and triple-cycle. They all work on the same principle, the only difference being the way the cam is shaped and the number of the cams and switches.

If the timer does not advance, check the AC voltage, 110 V, at the timer motor terminals. No voltage indicates switch contact trouble or bad connections. However, if you do measure 110 VAC, the timer motor is bad or jammed. Inspect the timer shaft and knob to make sure they are not rubbing against something causing the motor to stall. If nothing is binding, remove the timer motor leads and check the motor resistance. A good timer motor will probably measure around one thousand ohms. If you are still not sure, connect 110 VAC directly to the timer motor terminals with a test lead (see section 2-1). You should hear the timer operate and see the shaft move. Replace the motor if nothing happens.

Figure 8-2 shows a typical timer sequence chart. Even the most complex dryer timers are relatively simple and easy to understand. The left side vertical column shows the four timer switches: master contact, heat contact, timer motor contact, and extra-dry contact. This timer has three types of operation:

1. Unheated air cycle
2. Automatic cycle controlled thermostatically
3. Conventional timed cycle

The user selects the cycle on the front console.

Here's how to read the chart for the air operation. The master switch is closed as well as the timer motor switch for the first twenty-three minutes. The heat contact and the extra-dry contact are off during this same period. After twenty-three minutes, the master switch opens shutting down power to all functions. As far as the dryer operation is concerned, the drum and blower will rotate for twenty-three minutes and then stop. The heating device will never be activated. This is the simplest type of cycle.

Now refer to the middle cycle on the chart marked ''Automatic.'' At the beginning of this cycle the master switch, the heat switch, and the extra-dry switch are closed. The timer motor switch is open. During this first period the timer motor is controlled by a thermostat. After seventeen minutes of actual timer operation (real time is longer because the automatic function has kept the timer motor off much of the

DRYER TIMER SEQUENCE					
SWITCH	CYCLE				
	UNHEATED AIR	OFF	AUTO	OFF	TIMED	OFF
MASTER	23		24		73	
HEAT			19		68	
TIMER MOTOR	23		7		73	
EXTRA DRY			14			

Figure 8-2. Typical dryer timer sequence chart.

cycle), the timer motor contact is closed. Now the timer is in direct control for the remainder of the cycle. After nineteen minutes the heat contact will open. The remaining five minutes will be used to cool down the clothes with unheated air. The timer will turn off after twenty-four minutes of the timer motor time.

The third type of cycle shown in Figure 8-2 is the timed cycle. This is straightforward operation where the timer controls the cycle with exact time. At the start of this cycle the master contact, the heat contact, and the timer motor contact are activated. The extra-dry contact is open. All timer switches will remain the same for sixty-eight minutes, then the heat switch will open giving the clothes five minutes for cooldown. At this time the timed cycle is finished and all timer contacts are open.

When you suspect timer switch trouble, measure across the switch terminals with a VOM set on R × 1 scale (unplug power cord). A closed switch will measure 0 ohms, an open switch infinity or some resistance if there is a parallel path. You can turn the timer by hand while monitoring the VOM. A good switch will open and close according to the sequence shown on the timer chart.

Another technique to use with suspected bad timer contacts is to jumper a clip lead across them. If the load they are switching starts you can be sure they are not closing properly. For instance, suppose the timer is advancing properly but nothing is operating in the dryer. Shorting the master timer contact with a clip lead will immediately restore motor and heating operation if the switch contact is defective. You can do this with any timer contact. Just be sure the contact is supposed to be closed at the time you jumper it.

8-3. How to Repair the Heating System

The heating is accomplished with resistance wire coils mounted in the heating duct. The wattages and numbers of coils vary depending on the dryer model. A simple dryer may have just one heating coil, while a more sophisticated one could have as many as three. Due to the high wattage needed for resistance wire heating, 220 VAC is needed for power.

Figure 8-3 shows the heating element circuitry for a three-heat model. The user can select the amount of heat with a three-position

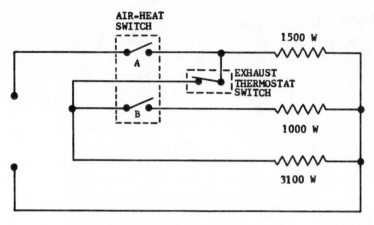

Figure 8-3. Three-heat element circuitry.

air-heat switch. The switch positions are marked "air," "normal," and "delicate." When the switch is on "air" all electrical paths to the heating elements are opened. Unheated air will be blown over the clothes.

The air-heat switch when placed in the normal position will close contacts A and B. Now all coils, 1,000W, 1,500W, 3,100W, will receive power. When the exhaust thermostat reaches the required temperature it will automatically open removing power to the 1,000W and 3,100W coils. The 1,500W coil will still receive power. The exhaust thermostat will temperature cycle the parallel 1,000W and 3,100W coils.

Moving the air-heat switch to "delicate" will close contact A and open contact B. The dryer now operates on 1,500W and 3,100W coils. The exhaust thermostat will now temperature cycle only the 3,100W coil.

If you are experiencing heating element problems, just measure the voltage, 220 VAC, across each heating element. When voltage is present and the coil is not heating, it probably has opened up. After turning off the power you can confirm your suspicions by measuring the continuity of the coil. If it's good it will measure a low resistance. Be sure you are not measuring two coils in parallel. If one coil is burned out, you will still obtain a good ohmmeter measurement. You'll have to pull off the connecting leads to isolate each coil. After you have found the bad coil it's usually a simple matter to replace it with a new unit.

8-4. Checking Thermostatic Controls

Dryer thermostats are used to change heat energy into mechanical motion, which activates switch contacts. There are three basic applications for dryer thermostats:

1. Operating thermostat
2. Safety thermostat
3. Cooldown thermostats

The operating thermostat is used to control the air temperature in the dryer. This can be done in a number of ways. Sometimes just one fixed temperature thermostat is all that's used. In other models a number of fixed temperature thermostats are selected by the console temperature switch. Some dryers are equipped with an adjustable thermostat that can be varied for a range of temperatures.

Safety thermostats are placed in strategic locations to sense an overheated condition. Most of these devices are normally closed and will open with an abnormal temperature rise. A common problem occurs when the lint filter is not cleaned periodically. The exhaust air cannot vent properly creating a high heat condition in the dryer. The safety thermostat will sense this condition and turn off the heating device.

Cooldown thermostats provide a path for motor power during the cooldown period. At the end of the heat cycle these thermostats operate until the temperature decreases to a fixed point. Then they will open to end the cycle.

Figure 8-4A shows a typical fixed temperature thermostat switch. These devices use a bimetallic disc that will snap from one position to

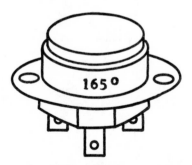

Figure 8-4A. Fixed thermostat.

another when heated, causing a switch to be activated. The switches can be single pole-single throw, single pole-double throw, normally closed, or normally open. They operate at various temperatures, which are generally stamped on the thermostat case. A typical operating thermostat will activate at 155° and deactivate at 140°.

Figure 8-4B shows a typical thermostatic-controlled circuit and how to check it with a VOM. Simply connect the VOM set on 220 VAC across the thermostat terminals and run the dryer through its cycle. When the timer switch and the centrifugal switch close, the heating element should heat. Both operating and safety thermostats should be closed with neither VOM measuring voltage. If either VOM measures 220 VAC, the associated thermostat is defective. As soon as the operating thermostat reaches its specified temperature, it will snap open and shut off power to the heater; VOM A will measure 220 VAC. When the temperature cools down the thermostat will automatically close again; the VOM voltage will drop to zero and the heater will be on again.

If you don't want to measure voltage in a 220 VAC circuit, you can use another method to check fixed temperature thermostats. With no power applied to the dryer, measure the resistance across the thermostat's contacts. It will measure 0 ohms on a normally closed contact and infinite ohms on a normally open contact. Aim a heat lamp

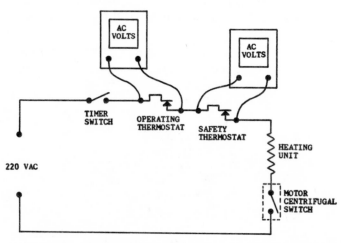

Figure 8-4B. Checking a thermostatic-controlled circuit with a VOM.

or hair dryer at the thermostat while the VOM is connected. As soon as the thermostat reaches its operating temperature you'll hear a small click and see the VOM pointer move to the other end of the ohms scale. If this action does not occur, replace the thermostat with a new one with the same specifications.

Adjustable thermostats consist of a bulb, a capillary tube, a bellows, a switch, and a control shaft. Here's the way it operates. The bulb is located in the exhaust air stream. When the temperature reaches a specified level a liquid inside of the bulb will vaporize creating pressure inside of the capillary tube. This pressure will be transferred to the bellows, which activates the switch. When the temperature returns to normal the gas changes back to a liquid state relieving the pressure. The switch contacts will return to normal position.

The adjusting shaft makes it possible to vary the expansion of the bellows so various temperatures can be made to activate the switch. You can test the adjustable thermostat using the resistance method described for fixed thermostats. It's helpful to monitor the exact temperature with a thermometer so you can verify that the thermostat is activating at various temperatures.

8-5. Understanding Electronic Dryer Controls

Some dryers do not depend on thermostat action for controlling drying temperature. Electronically controlled dryers depend on the moisture content of the clothes to control the heating unit. The electrical resistance of the wet clothes is monitored by sensing contacts within the drum. As the moisture content decreases to a minimum amount, the heating unit will be turned off by the electronic control circuitry.

Inside of the dryer drum are located sensor contacts that are touched by the wet clothes. Generally one of the contacts is grounded; the other is insulated and connected to the control input of the electronic unit. As long as dampness is present in the clothes a small current will be able to pass from the grounded sensor through the clothes to the insulated sensor and into the electronic control unit. When the current becomes too small because the clothes have become almost dry, the control unit will shut off the heating unit and will transfer motor control to a conventional cooldown thermostat.

When a defect occurs in the dryer and you suspect the electronic

control mechanism, it's a good policy to check all the other pos-
sibilities first. Electronic control units are reliable and other parts will
usually fail before them. Check out the sensors and the connection
system between the drum assembly and the control unit with ohmmeter
measurements. Be sure to check the brush connections or whatever
mechanism is used to transfer the continuity of the inside sensor to the
fixed outside connection.

Generally electronic controls operate a relay with multiple contacts. If the relay seems to be energizing correctly but the drum and/or
heating unit are not working, you can assume that the electronic control is okay but one of the relay contacts is probably defective. If a clip
lead jumpered across a "closed" relay contact restores normal operations, you can be sure that contact is bad. Another check that can be
made on the relay is to operate it manually with no electrical power
applied. Monitor the contacts with an ohmmeter to check for correct
operation.

Figure 8-5 shows a simplified electronic control dryer circuit.
When the start button is momentarily pushed, current will find a path
from line 1 through the electronic control and relay coil to the neutral
line. Relay contacts A, B, and C will close. Contact A will run the
motor. Contact B will keep the relay coil energized as long as the
sensing mechanism detects dampness in the clothes. The heating unit

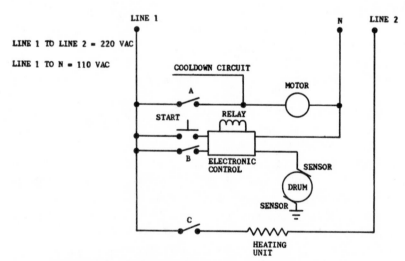

Figure 8-5. Simplified electronic control dryer circuit.

will be turned on with contact C. As soon as most of the moisture is removed from the clothes, the electronic control will de-energize the relay coil opening all contacts. Generally at this time another circuit controlled by a cooldown thermostat will keep the motor running until the clothes temperature drops below 120°.

8-6. Making Repairs on the Blower

The blower is necessary on automatic dryers to force air through the burner area and heater box into the drum area. Here the heated air absorbs moisture from the clothes and exhausts through the lint filter out of the machine. The main motor is responsible for moving the blower impeller. Some dryers have a single belt or a system of belts to transfer the motor power. In other models the blower is fastened directly to one end of the motor shaft.

Blower problems can usually be classified into three areas:

1. Mechanical failure of blower bearings and shaft.
2. A malfunction of the drive train between the motor and the blower.
3. A defective motor.

Since the blower impeller, bearings, and shaft are constantly operating in a heated environment, lubrication difficulties are often the cause of trouble. Many blower assemblies have provision for oiling but generally are inside of the machine and hard to reach (see Figure 8-6). When the lubrication dries out the bearings and shaft will become sticky and eventually scored. A trouble of this nature will usually become evident to the user by a shrill, squeaking noise when the machine is running. Often the noise will decrease or stop after the machine has run for a few minutes. If the bearings or shaft have not been scored, a few drops of oil in the bearing cavity should take care of the problem. When lubrication is ignored the shaft will eventually freeze to the bearings and stop dryer operations, usually overloading the motor causing the fuse or a thermostat to open. Once the blower shaft scores, the only solution is to replace it along with the bearing assembly.

Sometimes a malfunctioning blower is simply due to problems with the belt between the motor and the blower pulley. If the belt

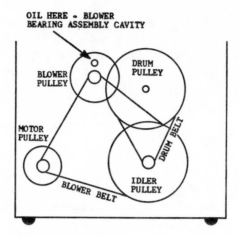

Figure 8-6. Typical blower lubrication area.

stretches or breaks the blower will operate slowly or not at all. Tightening or replacing the belt will often cure the problem. Make sure to turn the blower impeller shaft by hand before correcting a belt malfunction. If the impeller turns hard or not at all, a tightened or new belt will not cure the difficulty. The shaft has seized up and will have to be freed either by lubrication or replacement. When there is more than one pulley associated with the blower belt, turn each pulley by hand to make certain they are all turning freely.

Blowers that are connected directly to the motor shaft are less likely to fail than the ones with belt drive. Usually a blower failure will be the result of a motor failure. Section 8-7 will discuss motor problems in automatic dryers.

8-7. Testing the Motor

Most dryer motors are single-speed, split-phase, 1/4 to 1/3 horsepower motors (see section 6-5). The most unusual part of a dryer motor is the centrifugal switch contacts. All split-phase motors have a centrifugal switch that drops out the contacts on the start winding after full speed is attained. However, dryer motors have another set of contacts on the centrifugal switch that close the circuit to the heating unit. This is a safety precaution to ensure that the drum is rotating when the burner is started. If these contacts do not close properly, the heating unit will never start. Figure 8-7 shows a simplified dryer motor and heating unit schematic.

If the dryer is not heating but the drum is turning, one of the problems could be the centrifugal switch contacts for the heating unit. To check their operation, simply hook the VOM across the contacts (A and B in Figure 8-7). If you're working with an electric dryer set the VOM for 220 VAC; for a gas dryer 110 VAC. Clip the leads on the motor terminals before applying power. If the centrifugal switch heat unit contacts are operating properly, the VOM will measure line voltage for only a second or two after the motor starts; then it will drop to zero. You can be sure contacts are defective if the VOM reads constant voltage when the motor is running.

Many dryer motors have the centrifugal switch contacts mounted outside the motor on the motor shaft. This unit can be replaced easily by pulling off the connecting wires and removing a couple of mounting screws. Before removing the external centrifugal switch, observe how the actuating arm of the switch is located. After replacing be sure that no wires interfere with any of the movable parts associated with the centrifugal switch. Some dryer motors have the centrifugal switch inside of the motor housing. These contacts can also be replaced but require removal and disassembly of the motor (see section 6-5).

Many dryer motors using an external centrifugal switch are unrepairable. The case sections are welded or cemented together making it

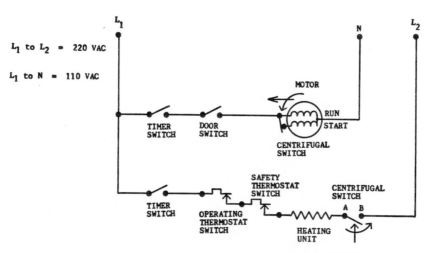

Figure 8-7. Dryer motor and heating unit schematic.

impossible to service. Most dryer motors are easy to reach for re-placement. Depending on the model you are working on, removing either the front or rear panel will give you access to the motor. Re-moval of the belt or belts, connecting wires, and the motor mounting screws or clips will generally be all that's necessary to pull the motor out. If the blower impeller is fastened to one end of the motor shaft, this will also have to be disconnected. Most impellers are screwed on the motor shaft with a left-hand thread. There's usually a flat on the motor shaft so that a wrench can hold it stationary while screwing off the impeller.

9

TROUBLESHOOTING ELECTRICAL AIR CONDITIONING PROBLEMS

All air conditioners operate on the same principle. In this chapter you'll find that it's easy to understand air conditioning operation and make electrical repairs.

We'll cover some of the ways to test the thermostat and the part it plays in overall air conditioning. Switches and relays are necessary air conditioning items that often become defective and have to be fixed. You'll find ways for diagnosing and repairing fan and blower malfunctions. Even though the compressor is a sealed unit in the refrigeration system, there are a number of electrical repairs that can be made to restore compressor operation. It's amazing how many refrigeration problems really turn out to be simple electrical defects.

Read on! It doesn't matter if you do not know the difference between a condenser or evaporator. If you want to repair a room air conditioner, read this chapter and you'll probably be successful.

9-1. Understanding How It Works

Figure 9-1A shows the block diagram of a typical air conditioning system. Freon refrigerant is pumped through the system on command of the thermostat. The refrigerant in the cooling area, evaporator, becomes a gas since it is under low pressure. The vaporizing gas absorbs heat from the surface of the evaporator. Air is then blown across the cold evaporator into the room producing the cooling effect. As the refrigerant gas is pumped under high pressure into the condenser by the compressor it turns into a liquid. When the refrigerant liquifies it gives up its heat and warms the radiating fins of the condenser coils. Air is blown across the condenser expelling the heat to the outside atmosphere.

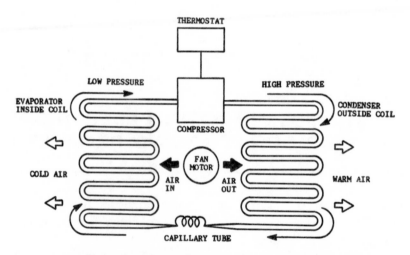

Figure 9-1A. Air conditioning system block diagram.

The liquid refrigerant must now return to the evaporator at low pressure so that it will be able to absorb heat again. It is forced through a small line, capillary tube, located between the condenser and the evaporator. The restriction of the capillary tube causes the refrigerant to reduce pressure, turning into a gas once more. The gas absorbs heat from the evaporator as before and the cooling cycle repeats.

Figure 9-1B illustrates a simplified electrical schematic of an air conditioner unit. The on-off switch controls both the fan and the compressor motors. The fan switch controls the fan motor for low or high

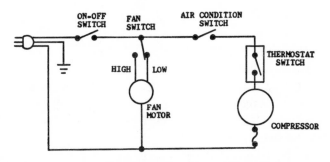

Figure 9-1B. Air conditioner electrical schematic.

speeds independent of compressor operation. Once the air conditioner switch is closed and the thermostat set, the compressor will begin operating until the desired temperature is reached. The fan motor runs at all times for maximum cooling effect.

9-2. Ways of Testing the Thermostat

The thermostat is simply an automatic switch controlled by temperature. It will close at a high temperature and open at a lower one. Most thermostat dials are calibrated with numbers to indicate degree of coldness desired.

A sensing bulb is the device used to determine air temperature. The bulb and its connecting copper tube are filled with refrigerant gas and positioned in the return air stream. Most bulbs are located in the intake port of the evaporator fan or in front of the evaporator. When the gas inside of the sensing unit warms, it expands pushing a bellows inside of the thermostat, which in turn closes the switch starting the compressor motor. Eventually the cooled air passing over the sensing bulb will reduce the sensor pressure allowing the bellows to contract and open the thermostat compressor switch. This cycle will repeat over and over to maintain selected air temperature.

You can check the sensing bulb operation with your body heat or with a container of cracked ice. Holding the bulb between your finger and thumb should raise the bulb temperature to activate the thermostat switch and compressor motor. Or inserting the bulb into cracked ice should immediately turn off the compressor.

Many thermostat problems can be checked by visual observation.

Disconnect power and remove the thermostat cover. Turn the dial while watching the switch contacts. The contacts should move, make contact, and break contact during the dial rotation. Most thermostat contacts can be adjusted. Generally turning a screw is all that's necessary to make the adjustment. If the contacts are not closing, turn the adjusting screw in the direction that allows the contacts to close. If they are not opening, turn the screw in the opposite direction. Once you get the contacts making and breaking, operate the air conditioner and readjust the contacts for desired operation.

If the contacts seem to be closing but the compressor does not operate, clip a jumper lead across the thermostat terminals. The compressor will immediately start if the thermostat is faulty. Don't try to file the contacts since they are coated with a precious metal for a good contact surface. Filing will remove the contact area causing them to arc and wear out prematurely. Most thermostats can be replaced quite easily just by removing a couple of screws. Make certain that the sensing bulb is mounted exactly as in the original position. If the sensor is too close to the evaporator coils it will "read" coil temperature instead of air temperature.

9-3. Checking Electrical Controls

The common controls on air conditioners are the thermostat, fan-air condition switch, and starting relay. The thermostat has already been covered in the previous section. The fan-air condition switch is often a rotary wafer switch that controls the following action:

Position	Operation
1	Off
2	Fan, high speed
3	Cool, high speed
4	Cool, low speed
5	Fan, low speed

Figure 9-3A shows the symbol for the previous switch. Notice the fan is on for every position except the off position. The compressor is activated in positions 3 and 4 through the thermostat. Many switch problems can be determined just by looking, listening, and feeling. When the switch is not a sealed unit a visual inspection will often show

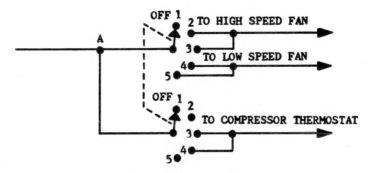

Figure 9-3A.   Fan-air condition switch.

defective contacts. Many times a defective switch will sound peculiar or feel tight, loose, or spongy when it's actuated.

A simple clip-lead test will verify switch troubles. Clip one side of a well-insulated, heavy-current clip lead to point A as shown in Figure 9-3A. Jumper the portion of the switch that doesn't work. For instance, suppose the fan never operates in the high-speed positions. Touching terminals 2 and 3 of the upper rotary switch with the other end of the clip lead will start the high-speed fan if the switch is defective. If the compressor is not working, simply touch terminals 3 and 4 on the lower rotary switch with the free end of the clip lead. The compressor will start if the switch is bad. Make sure the thermostat is set high enough to require air conditioning.

Some air conditioners use a voltage relay to replace the centrifugal starting switch in conventional motors (see section 6-5). The compressor motor in air conditioning units is sealed into the system so an external method is used to switch out the motor start winding when run speed is reached. Figure 9-3B shows the schematic of a voltage relay and how it controls the motor starting winding. The relay contacts are closed at start time, completing the start winding circuit through the capacitor. When the motor speed increases, the voltage dropped across it also increases. Since the relay coil is in parallel with the start winding it will pull due to the increased voltage. The start winding will be switched out of the circuit. The motor will continue running just using the run winding.

The start relay can be checked with a VOM ohmmeter for coil continuity and contact operation. Whenever the compressor won't run or the overload protector kicks out shortly after starting, the relay may

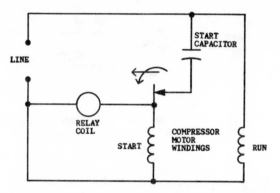

Figure 9-3B. Voltage relay and motor schematic.

be defective. You can measure the AC voltage across the compressor start terminals with the thermostat set for air conditioning. If you don't measure any voltage, the starter relay is probably not making contact.

9-4. Diagnosing and Fixing Blower Troubles

The fan motors in air conditioning units are often shaded-pole motors that start without using an extra start winding. This type of motor can be used because the fan load is very light at start-up time. Two- and three-speed fan motors are also common in many air conditioning units.

Figure 9-4 shows the schematic of a two-speed fan motor. Speed control is achieved by adding windings in the motor. Some motors have a capacitor across the input line. This capacitor tends to compensate for the inductance of the motor windings and improves the motor operation.

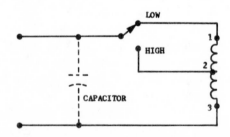

Figure 9-4. Two-speed fan motor schematic.

If the fan doesn't run, make sure that the fan motor is receiving voltage before you start working on the fan. An ACV VOM check between fan terminals 1 and 3 (low) and 2 and 3 (high) (Figure 9-4) will confirm whether the voltage is present. Absence of voltage probably indicates a bad switch. When voltage is present on the fan motor terminals but the fan is not running, the fan itself is probably at fault. Disconnect the fan motor terminals from the circuit and measure the resistance of the motor windings. A low resistance should be measured from terminals 1 and 2, and 2 and 3. A slightly higher resistance should be indicated from terminal 1 and 3. An infinite resistance should be measured from any motor terminal to its metal case. A resistance reading here indicates windings shorted to the case and the need for a replacement motor.

A common problem with fan motors is dirt and lack of lubrication. Try spinning the fan by hand. If it moves hard or not at all, the bearings are probably frozen. Often this condition can be repaired without replacing the motor. If the motor has oil cups or felt pads at the shaft ends, try oiling with a lightweight machine oil. Also oil the shaft lightly in such a manner that the oil will run onto the bearing surfaces. With any luck the shaft will loosen. Rotate the shaft manually until it spins freely. In stubborn cases the motor will probably have to be disassembled. Occasionally the blower will not turn because it's touching the housing or something is jamming the impeller. This is readily apparent by spinning the fan manually.

Capacitors sometimes go bad. Most fan motors will start even if the capacitor has opened. However, it won't start or run with the zip that it had normally. If the capacitor shorts it will blow the house fuse or trip a circuit breaker. In this case, disconnect one end of the capacitor and try operating the fan again. If the fan runs, the capacitor is the culprit. A shorted capacitor will measure a low resistance between one terminal and the other when it is disconnected from the circuit. An open capacitor will measure an infinite resistance across its terminals with no momentary resistance reading. The best way to check a capacitor is to replace it with an identical unit.

9-5. Repairs for the Compressor Unit

When you have a compressor problem, make sure it really is a compressor problem. If the fan and the compressor are both not operating at any switch position, chances are the air conditioner is not getting line voltage. Check the voltage at the outlet with the VOM set on AC volts. Many air conditioners operate on 220 VAC, so be sure the VOM is set on a high enough voltage range. Figure 9-5A shows how outlets can be checked with a VOM. VOM A and B will check 110 VAC if voltage is present. VOM C and E should also measure 110 VAC in the 220 VAC outlet. However VOM D will read 220 VAC. If the voltage is missing at the outlet, the circuit breaker or fuse for that circuit has probably opened in the entrance box. Check the current requirements of the air conditioner and the size of the fuse or circuit breaker that is protecting the circuit. If the air conditioner circuit is protected with conventional fast-blow fuses, it could be possible to blow the fuses under normal conditions. All motors, and especially ones like the compressor motor that start at full load, draw much more current to start than they use while running. High line voltage or wear and tear on the motor could cause starting current to be higher than normal. Or if a fuse has been operating at its rated current for any length of time it will often die from thermal stress. A simple remedy for this kind of prob- lem is to replace the fast-blow fuse with a delay fuse or slow-blow type. The current ratings are identical but the delay fuse will absorb an overload for a short time before popping. Circuit breakers have this delay feature built into their construction. However, circuit breakers sometimes wear out, too.

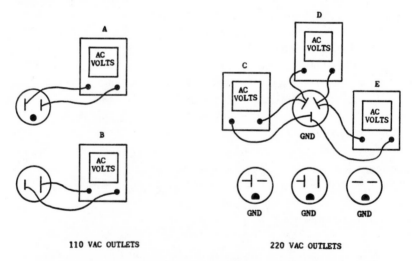

Figure 9-5A. Checking an outlet for AC voltage using a VOM.

Suppose line voltage is present and the fan operates okay but as soon as the compressor motor tries to start, it immediately quits. It's probably tripping the compressor overload protector. The starting capacitor is probably defective. The capacitor is generally a large, tubular capacitor located near the compressor. The simplest way to check the capacitor is to substitute it with a new, identical unit. Remove the wires from the old capacitor and temporarily attach them to the new capacitor terminals. Start up the air conditioner and if normal operation returns you can be sure the culprit was the starting capacitor.

Many compressors also have a run capacitor electrically located between the run and start windings. If this capacitor shorts out, the compressor will draw excessive current and trip the overload protector. A shorted capacitor will measure a small resistance across its disconnected terminals with a VOM ohmmeter. It's a good idea to substitute this capacitor with an identical unit to eliminate any possibility that it is defective.

Figure 9-5B shows the schematic of the windings of a typical compressor motor. Generally there are three terminals on the outside of the compressor housing. One terminal is the common connection between the run and start windings. The other terminals are for the start and run windings. When the compressor is working properly, line voltage should be measured between the common and the run terminals with a VOM. Line voltage will also be measured between the common and start terminals until the motor reaches running speed. Then the starter relay will open, stopping the current to the start winding. Some compressors do not have a start relay and always have voltage between the start windings and the common terminal.

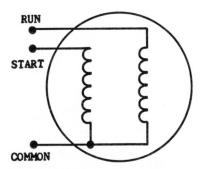

Figure 9-5B. Compressor motor schematic diagram.

If voltage is present but the compressor is not running or is tripping the overload protector, you can make an ohmmeter check on the compressor motor windings. Disconnect the wires from the compressor terminals so there are no parallel paths. Identify them so they can be replaced in the right place. A small resistance should be measured between the common terminal and the start terminals, usually about one ohm. The resistance between the common and the run windings should be less resistance. Many VOM ohmmeters are not sensitive enough to distinguish the slight difference in start and run winding resistance. There should be infinite resistance between each of the terminals and the compressor metal housing. Any resistance indicates a shorted winding and a compressor replacement will be necessary. Good ohmmeter readings do not always mean the compressor is okay. However, open windings and shorts to the case can be found reliably using the ohmmeter.

10

PRACTICAL WAYS TO REPAIR THE AUTOMOBILE IGNITION SYSTEM

This chapter covers the automobile ignition system, how you can repair it, and how the ignition switch, coil, distributor, spark plugs, points, and capacitor function in perfect harmony with each other to allow the engine to run. You will learn some things you should know about spark plug replacement and maintenance. The distributor is really a pretty simple device that can be repaired quite easily without a lot of expensive equipment. What about points and condensers? Here you will find some handy tips on checking, setting, and replacing these important ignition devices.

If you have ever been stranded in the middle of nowhere with an automobile electrical problem, you probably would have given an eyetooth to get that car rolling again. A lot of electrical breakdowns can be repaired temporarily if you know a few checks to make. A section of this chapter deals completely with emergency ignition repairs. You'll read about electronic ignition and see how easy it is to understand when you know how a conventional system operates.

10-1. Guide to Proper Operation

Turn the ignition key and the automobile engine will come to life. This simple act starts a chain of events that electrically starts and controls the engine operation. In this section let's follow the electrical sequence of a typical automobile ignition system. Refer to Figure 10-1.

As soon as the ignition switch is placed in the start position, current will find a series path from the negative post of the battery to ground. From ground it will flow through the points and the primary winding of the coil. It will continue through the coil primary to the ignition switch, then back to the positive battery terminal. At the same time, the starting motor has received power and is cranking the engine causing the points to close and open. Also the rotor switch is being turned around inside the distributor.

When the points open, the capacitor (condenser) across them will charge up. A magnetic field had been built up around the primary winding as current had flowed through it. Now the open points will stop current flow causing the magnetic field to collapse. The collapsing magnetic field will cut across the secondary coil winding, which has thousands of turns of wire. This action will induce a very high voltage, 20,000 volts or more, in the secondary winding.

The secondary coil winding is connected to the distributor center terminal with a high-tension wire. The high voltage energy will be transferred from the center terminal through the rotor to a spark plug. Remember, the rotor is turning so each spark plug will obtain a burst of high voltage during one rotor revolution. At the spark plug the voltage

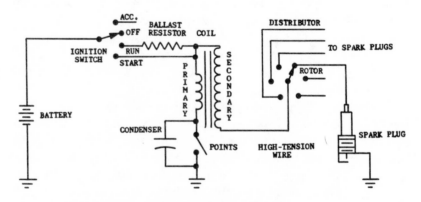

Figure 10-1. Simplified automobile ignition circuit.

will be high enough to jump the gap between the center electrode and the grounded electrode. Since this end of the plug is inside one of the firing chambers of the engine, the spark will ignite gases and produce the explosion that causes internal combustion energy.

When the engine starts, the ignition switch is moved to the run position. The previous ignition action will continue except that a ballast resistor is now in series with the primary circuit. The resistor drops the primary voltage to about one-half the battery voltage at low engine speeds. When the engine speeds up, less current is drawn through the resistor, decreasing its resistance. Now more primary voltage is available at the coil. The resistor reduces coil current during low engine speeds and prolongs point contact life. Many ballast resistors are mounted on the firewall. Some cars use a resistance cable instead of a resistor between the ignition switch and the primary coil terminal.

The capacitor mounted across the points does several things for correct ignition operation. First, it absorbs the current that would like to continue through the point contacts as they are opening. This would cause point arcing and drastically reduce point life. It would also decrease the amount of energy left for the high-voltage buildup. The energy stored in the capacitor will also be available for increasing primary coil voltage when the magnetic field collapses.

10-2. Spark Plug Checking Made Easy

Most spark plugs last between 10,000 to 12,000 miles, depending on how the car is driven and engine design. Manufacturers generally recommend removing, cleaning, testing, and regapping every 5,000 miles. If your car is designed to make plug removal difficult, you're probably better off just replacing the set of plugs around every 8,000 miles. Driving in city stop-and-go traffic will generally cause the plugs to eventually foul with unburned gas causing rough idle. Worn spark plugs need more voltage to fire them than new ones. This is most noticeable when accelerating since the spark plug voltage requirement is greatest during hard acceleration. If the engine sputters and misfires during acceleration but runs smoothly otherwise, it could be telling you it needs a spark plug change.

If your engine is missing and running rough, some of the prime suspects to check are the spark plugs. An easy way to make a quick check on a spark plug to find out if it is firing is to short it out with a screwdriver blade (Figure 10-2A). Using a screwdriver with a well-insulated handle, touch the tip to the engine block and the screwdriver shank to the top of the plug. This will ground out the plug and cause that cylinder to stop firing. The engine will miss and run rough. If you can't detect any difference in engine performance while making this test, you can be sure that the plug is not firing.

When in doubt as to whether a plug is receiving high voltage, simply remove the high-tension wire from the plug and position it about 1/2 inch from the engine. Start the engine and you should see a strong spark jump the gap between the end of the spark plug cable and the block. If each of the spark plug cables passes this test, you know that high voltage is available at the input side of the plugs. The plug may not be firing in the engine because it is defective but the voltage is available.

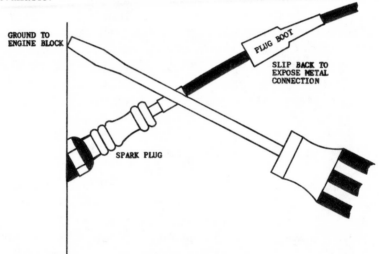

Figure 10-2A. Shorting a spark plug with a screwdriver blade.

After removal of the spark plugs from the engine, inspect them carefully. Generally it's a good idea to remove just one plug at a time to eliminate the chances of crossing the plug wires. If the plug is in good shape it will show little signs of wear. Normal deposits will be light tan or gray. The gap will be close to the recommended setting and the electrodes will have sharply defined edges (Figures 10-2B). Check the porcelain insulator for cracks, especially where it enters the metal

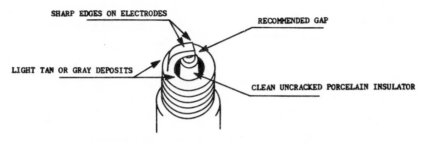

Figure 10-2B. Normal spark plug condition.

shell. If the plugs show no excessive electrode wear, fouling, bridging, or other abnormal condition, they can be cleaned, gapped and used again. Scrape away the deposits inside of the plug with a sharp, thin scratch awl being careful not to stress the ceramic around the center electrode. File the electrodes so they have sharp edges again and resemble a new plug (Figure 10-2C). Reset the gap with a wire-type feeler gauge to the proper specifications. If you're replacing the plugs, be sure to also check the gap on the new ones.

You can check the continuity of resistor-type plugs with a VOM ohmmeter. A good plug will measure between 5,000 and 10,000 ohms between the top terminal and the center electrode. This is not a critical resistance measurement, but should remain constant when the plug is shaken or tapped. An infinite reading indicates a bad plug. Measuring between the center and the outside electrode should always measure infinity.

Coating the plug threads with a little graphite grease will make them a lot easier to remove the next time they're checked. Screw the plugs in finger tight, then wrench tight 1/2 to 3/4 turn. Don't put any

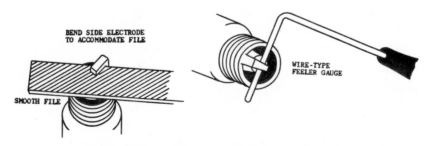

Figure 10-2C. Filing and gapping a spark plug.

side pressure on the spark plug wrench since it's very easy to crack the ceramic insulator. A good preventative maintenance program will usually eliminate any major spark plug problems.

10-3. Servicing Defective Distributors

The distributor is the source of a majority of ignition problems in the automobile. High-voltage switching and mechanical motion are the reasons for many failures. The major electrical components in the distributor are:

1. Points (switch contacts)
2. Condenser (capacitor)
3. Rotor (rotary switch armature)
4. Distributor cover (rotary switch contacts)

The points are used to open and close the coil primary circuit, which is necessary to produce high voltage at the spark plugs. The condenser absorbs electrical energy when the points open and aid coil operation. The rotor and distributor cover are in the secondary high-voltage circuit. The rotor transfers the high voltage sequentially from the distributor cover center terminal to each of the contacts around the perimeter.

During engine operation the points are the source of most distributor troubles; the switch contacts are opened and closed thousands of times each minute. The points can arc causing burning and pitting. Once this happens the points should be replaced and the condition causing excessive point deterioration repaired. Burning contacts result from four separate causes:

1. High primary voltage
2. Oil or grease contamination
3. Defective capacitor
4. Wrong point adjustment

High primary voltage causes excessive current to flow through the points. This is a result of a defective or misadjusted voltage regulator. If the ballast resistor has been shorted out for any reason, this, too, would cause heavy point current. Crankcase vapors or overoiling of the distributor can deposit oil on the contacts. If the distributor cam has been excessively greased or lubricated with the wrong type of lubricant, this can be thrown on the point surfaces. Points that open too little because of misadjustment or rubbing block wear will burn. A defective capacitor or any high-resistance connections in the capacitor circuit will cause premature point failure.

Pitting of the points results from the transfer of material from one contact to the other. All points generally pit to some degree, but an excessive amount is not normal. Abnormal pitting is caused by:

1. Small point gap
2. High primary voltage
3. Wrong or defective capacitor

When points are replaced be sure to align the contacts so they mate evenly. Bending the stationary contact slightly is all that's necessary for perfect point alignment. Set the point gap with a clean feeler gauge to the proper setting, as shown in Figure 10-3. When the gap is correct you'll feel a very slight drag on the gauge as you remove it from between the point contacts. You are generally better off to replace points rather than trying to repair them. Be sure to coat the cam with a light coating of special grease before buttoning up the distributor. This will keep the rubbing block from wearing out prematurely. This grease will not fly out and contaminate the points. Unless a capacitor checker is available, you might as well replace the capacitor when you change the points. Capacitors generally last much longer than points, but good preventive maintenance dictates a capacitor change when you are in doubt as to its condition.

Always time the engine after the points have been replaced. Clean the timing marks thoroughly with a rag, marking the specified timing lines and pointer with white chalk. Insert a timing light between the number one cylinder spark plug and its high-tension cable (or follow

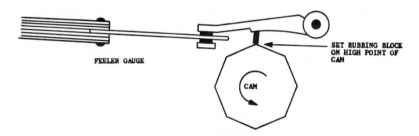

Figure 10-3. Setting the point gap with a feeler gauge.

timing light directions). Loosen the distributor mounting nut and pull off the distributor vacuum hose. Start the car and allow it to warm up. Aim the timing light at the timing marks while turning the distributor housing. *Caution: Keep away from the fan blade!* When the timing marks match each other, tighten up the distributor hold-down and the job is finished.

When you're having trouble in the high-voltage secondary circuit, prime suspects to check are the distributor cap, the rotor, and the high-tension leads. Pull off the distributor cap and carefully inspect the inside and outside for cracks or burnt carbon paths in the plastic. If you have to replace the cap, pull off the high-tension wires carefully one at a time and insert into the new cap. This will prevent you from mixing up the wires. Be sure to identify the cap terminals correctly before inserting the cables. If the distributor cap is good, clean it up and remove any accumulation of deposits from the inside cap terminals with a screwdriver blade.

Check the rotor for deterioration. Especially notice the condition of the tip for corrosion or burning. Also check the spring contact, which mates with the center terminal of the distributor cap, for condition and tension. When replacing the rotor be sure that the new rotor is identical to the old one. Many replacement rotors are not the same as the original one and can cause all sorts of problems.

Check the high-tension cable connections at the coil, distributor, and spark plugs. The terminals should be fully seated with good electrical contact between the conductor and cable terminals. The rubber nipples should fit tightly on the coil and distributor towers. The spark plug covers should fit tightly around the porcelain insulators. The high-voltage cables as well as the distributor and coil cases should be cleaned with a solvent and wiped dry. The cables should not be brittle or cracked.

Here's a good method to test the high-tension cables for leakage. Disconnect a high-voltage cable from a spark plug. Position the cable so it will not arc to any grounded point. Ground one end of a clip lead to the engine. Start the engine. Move the probe end of the test lead along the length of the high-tension lead. A spark will jump to the test probe from any cracks or punctures in the wire. If this wire checks okay, make the same test on the high-voltage wire between the coil and distributor center post. Each spark plug wire should be checked with this test.

Continuity of high-tension cables can be checked with the VOM ohmmeter. Most cables conduct the high voltage with some type of graphite-impregnated, center-core conductor. This resistance cable prevents radio and television interference from your car's engine. Lift the distributor cap. Fasten one ohmmeter lead to an electrode terminal inside of the cap. Attach the other ohmmeter lead to the spark plug end of the cable. Different resistance cables have different resistances. Most spark plug cables will measure between 20,000 and 30,000 ohms. Any appreciable resistance above 30,000 ohms indicates that a new cable is needed. It's a good idea to wiggle the cable gently while making the ohmmeter test. A good cable will keep its resistance value constant regardless of the flexing. To check the center distributor cable connect the ohmmeter probe to the inside center terminal of the cap. Use the other ohmmeter lead to probe the primary terminal on the coil. This test will measure the resistance of the center high-tension cable and the coil secondary resistance; a good test would indicate less than 25,000 ohms. If a significantly higher resistance is measured, determine whether the cable, the coil, or the connections are the culprit by further ohmmeter checking.

10-4. Remedy for Coil Problems

The ignition coil is generally one of the most reliable parts in the automobile electrical system. The coil is really a step-up transformer responsible for changing the primary car voltage, 6 or 12 VDC, into the secondary high voltage, 20,000 to 25,000 V, which is needed to create a spark at the spark plug electrodes.

If you cut a coil apart you will see a soft iron core in the center. Wound around the core is the secondary winding consisting of thousands of turns of fine copper wire. The primary winding has only a few hundred turns of heavier wire wound over the secondary winding. The windings and core are inserted into the coil case and surrounded with insulating oil or pitch and sealed. The windings are usually terminated with three connections in the coil case. The push-into center connection is for the high-voltage secondary lead. The other two connections are screw terminals for the primary circuit. One of these terminals is generally marked with the ungrounded, hot, battery terminal polarity. Generally two wires are fastened here. One wire provides high primary current during starting; the other wire provides normal primary current during running. The second primary terminal is connected directly to the points inside of the distributor.

When you suspect a defective coil here are some simple tests you can make. Pull off the center high-tension cable from the distributor cap. Try to be as gentle as possible when you do this so you don't rip the center conductor from its terminal. Position the end of the cable a quarter-inch or so away from a grounded part of the engine while operating the starter (Figure 10-4A). You will see a strong, bright, blue-colored spark jump the gap if the coil and associated circuits are okay. A weak or nonexistent spark indicates a defective coil or trouble in the primary ignition circuit.

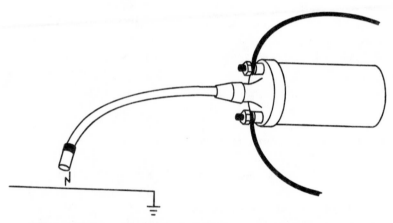

Figure 10-4A. Checking for spark on a coil's high-tension secondary lead.

If trouble is indicated, check the primary circuit with a VOM or 12V test bulb. Take off the distributor cap and make sure the points are open. If you're not sure stick a piece of thin cardboard between the point contacts. Turn on the ignition switch to the run position. Set the VOM for 12 VDC and connect one lead to ground. Refer to Figure 10-4B for the following tests. Measure the voltage at coil terminal A. Full battery voltage here indicates the battery, ignition switch, ballast resistor, and connecting wires are okay. Now measure the voltage at coil terminal B. Full voltage here indicates the coil primary is not open and probably okay. No voltage indicates a defective coil primary. If

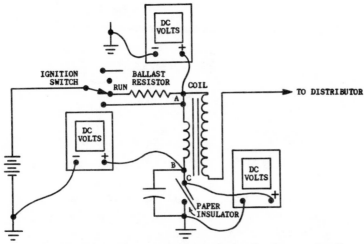

Figure 10-4B. Testing voltages in the coil primary circuit.

you have to replace the coil don't put excessive force on the primary connecting screws while tightening the nuts. They will break if tightened with too much force. If the voltage is normal at B, try measuring at point C. Voltage absence here indicates that the connecting wire from the coil to the distributor has opened.

When the primary voltage tests are okay try this technique. Position the distributor end of the coil secondary HV cable close to a grounded metal part. Now short the points with a clip lead or screwdriver blade while watching the end of the HV cable. A spark should jump from the HV lead to ground each time the clip lead or screwdriver blade is removed. If the spark is evident with this test but not before, you can bet the points are defective and need replacement and adjusting.

Be sure to observe coil polarity when replacing the coil. When the wires to the two primary terminals are interchanged, the high voltage needed to fire the spark plugs will be considerably higher than specifications. Most coil primary terminals are marked as shown in Figure 10-4C. For all cars with a negative ground (negative battery terminal fastened to engine block) the ignition terminal on the coil is marked BAT. or +. The other primary terminal is marked DIST. or −. On positive ground systems the primary coil polarity is reversed.

Sometimes a coil will begin arcing at the high-voltage secondary terminal due to dirt or moisture accumulation. This will usually cause

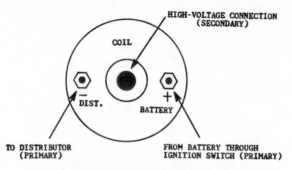

Figure 10-4C. Coil terminals designation.

the spark path to carbonize the plastic coil case and the rubber boot. You can identify this problem easily at night while watching the coil's high-voltage connection when the motor is running. The arcing will be evident. When arcing occurs it's very difficult to remove the carbonized path. Your best bet is to replace the coil and the rubber boot on the secondary connection.

10-5. Emergency Electrical Repairs

The best insurance for avoiding electrical breakdowns is a good preventive maintenance program for your car's electrical system. However, even the best-maintained cars occasionally fail and it's a great help to know how to restore operation so you can be on your way again.

A common problem is getting the car started. Suppose you turn the ignition key and nothing happens. Your battery could have died but the chances are that's not the trouble. A battery that is about to fail will often give warning signals such as slow or labored cranking, the turn signal flasher clicking slowly, the lights dimming excessively at idle, and so on. Even a very weak or almost dead battery will generally operate the starter solenoid and make a slight effort to crank the engine. Assuming that the battery has been functioning properly, what can be done to get the car started?

Try your horn and lights. If they operate normally your battery is okay. Suppose the horn, lights, or for that matter no electrical apparatus in the car is operating. The problem could be a high-resistance connection between the battery post and cable clamp. Occasionally a film of oxidation will completely insulate the battery terminal from the cable clamp. Wiggle both battery connections as well as the ground connection to the engine. Sometimes this is all that's necessary to break the oxide film. If the car starts you're on your way. When wiggling doesn't pay off, try shorting a screwdriver blade across the battery terminal and its cable clamp. Twist the screwdriver blade so that it digs into the lead making positive contact. If possible, try to start the engine while doing this operation. This will break the oxidation and the car will start like magic.

When the lights and horn do work but the car won't crank, the heavy cable from the battery to the solenoid switch or to the starter could have become loose or oxidized. Wiggle these connections back and forth. If the car starts don't forget about the bad connection because it will happen again unless it's properly fixed. See sections 11-2 and 11-5 for details.

Now suppose the car cranks all right but the engine won't run. One of the easiest ways to check the ignition circuit is to pull off a spark plug wire and hold it close to the engine block. Crank the engine and look for a spark. A good spark indicates the electrical system is probably okay. If it is an electrical problem the plugs would now be the prime suspects. Think about how long the plugs have been in the engine. If they have been in for ages this is probably the problem. Pull out three or four plugs (it's handy to carry a spark plug wrench in the car) and look them over. An extra-large gap or oil-fouled, gasoline-saturated electrodes could be the answer. Clean up the plugs as well as you can and reduce the plug gap (a paper clip makes a fine emergency feeler gauge). Often cleaning up just three or four plugs is all that's necessary to get the old buggy running again.

If there is no spark on the plug wire, work back into the ignition circuit. See if there's a spark from the high-tension lead of the coil to ground when it's removed from the distributor and positioned close to the engine. A good spark indicates a good ignition switch, coil, points and condenser. Maybe the distributor cap has popped loose or someone has stolen your rotor. It's always a good idea to keep a few spare ignition parts in your glove compartment: rotor, points, capacitor. If you know the position of the rotor in relation to the flat of the distributor shaft and have a roll of electrical tape and a paper clip, make a rotor. Wrap the end of the distributor shaft with lots of electrical tape to insulate the high voltage. Bend the paper clip into an L shape approximately the same length as the rotor. Tape it securely on the distributor shaft so there will be a high-voltage path from the distributor cap center terminal to the other contacts in the cap. This won't be the best rotor in the world but it will probably get you home.

No spark at the coil high-voltage lead could indicate ignition primary troubles. The most likely cause of trouble would be the points. Perhaps the rubbing block has worn down enough so that the points will never open. Reset the point gap using a makeshift feeler gauge; the thickness of a matchbook cover is just about right for the gap. In fact, while you have the matchbook cover out, you can use the striker area to clean up the point surfaces. If the points are excessively pitted or even welded together, the capacitor has probably opened. Separate and clean up the points as well as you can using whatever tools that are available. A fingernail file could work wonders. Here's where that spare capacitor would become valuable.

Suppose the point gap appears normal and there's no excessive pitting. Crank the starter so that the points are closed. With the ignition switch on, open and close the points by hand. You'll see a small spark across the points if primary voltage is there. If no spark is seen, try sliding and rubbing the point contacts against each other. This will often break the oxidation film on the contacts. If still no spark is seen, remove the coil primary lead or leads (the terminal that doesn't go to the distributor) and flash them across a ground. A spark will be evident if the coil wire is getting primary voltage. No spark could mean a defective ignition switch. If this is the case, clipping a wire from the ungrounded battery terminal to the coil primary terminal (the terminal that doesn't go to the distributor) will restore voltage. Don't operate the car for long in this condition since the ignition switch will not be able to shut off the engine anymore. In addition, the coil and points will be receiving excessive current since the ballast resistor has been bypassed.

Most of these tests and repairs are fairly quick and simple to make. Hopefully when you have emergency trouble with your car's electrical system one of the previous suggestions will pull you through the difficulty. The key to avoiding almost all emergency repairs is to keep your car's electrical system in tiptop shape with periodic inspections and maintenance. Anticipate some of the problems that could occur and carry enough tools and small replacement parts to do the job. If you have trouble, you'll be glad you have them.

10-6. Understanding Electronic Ignition

As well as the conventional ignition system performs, the electronic ignition system performs better. There are many variations of electronic ignition. At the writing of this book Chrysler Corporation is equipping all cars made in North America with electronic ignition. Chances are that all cars will eventually run using some form of electronic ignition.

Chrysler's electronic ignition system eliminates the points and condenser of conventional ignition. Since the points generally have the shortest lifespan of all ignition components, the electronic system eliminates maintenance and failure of the breaker points. Also, much undesirable exhaust emissions will be stopped because the ignition system will remain tuned for a longer period.

Most of the components in Chrysler's electronic ignition system appear to be identical to the standard breaker-point system. In fact, the distributor case and cap, rotor, coil, and spark plugs remain the same. The ballast resistor has been replaced with a dual ballast resistor with four connections. Inside the distributor the rubbing block and cam have been replaced with a gearlike component called the reluctor. A permanent magnet, coil and pole piece (all of which constitute the pickup unit) substitute for the points. An electronic control unit mounted on the firewall determines dwell and fires the primary circuit of the coil.

Figure 10-6A shows the pickup unit and reluctor of an electronic ignition system distributor. As the reluctor moves past the pole piece,

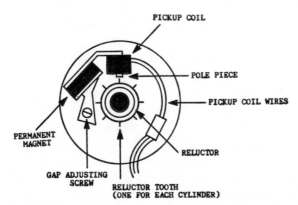

Figure 10-6A. Distributor of an electronic ignition system.

the magnetic field in the pickup coil is increased. When the reluctor passes the pole piece the magnetic field in the pickup coil decreases causing a negative-going voltage to trigger the electronic control unit and stops the switching transistor from conducting causing interruption of the primary coil current. The collapsing of the coil primary induces the high voltage in the coil secondary exactly as the standard ignition system. The reluctor and the pickup unit determine ignition timing while the control unit determines dwell.

The ballast resistor (Figure 10-6B) has a double role in electronic ignition operation. One resistor, the .55 ohm, does the same job as the ballast resistor in the conventional breaker-point system. The 5 ohm resistor limits current in the electronic control unit.

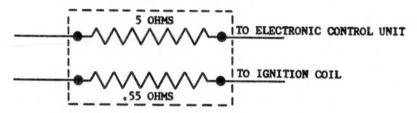

Figure 10-6B. Electronic ignition ballast resistor.

Figure 10-6C shows the circuit of an electronic ignition system. You can troubleshoot this system rather easily with just a VOM and a little common sense. Keep in mind that the control unit and the coil are probably the most reliable parts of the system and should be the last to require checking.

When you suspect ignition trouble, pull the coil center high-voltage cable from the distributor and hold about 3/16″ from a ground point. Crank the engine while watching for a spark. A good spark indicates no ignition trouble. A very weak spark or no spark means ignition trouble and further troubleshooting.

Turn off the ignition switch and pull the multi-wiring connector from the control unit. *Caution:* Always make certain the ignition switch is off when removing the connector to prevent damage to the control unit. After the connector is removed, turn the ignition switch to ''on'' and turn all accessories off. Measure the voltage at pin 1 cavity of the connector with a VOM. It should measure +12 VDC or very close to it. Lack of voltage indicates a bad ignition switch or wiring connections in that circuit (see Figure 10-6C).

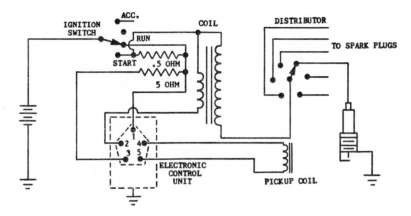

Figure 10-6C. Circuit of an electronic ignition system.

Next, measure the voltage at pin 2 cavity of the connector. Again, the voltage should read + 12 VDC or slightly less. A lack of voltage here would indicate an open .55 ohm ballast resistor, bad connection, or coil primary. Refer to Figure 10-6C and make an ohmmeter check of pin 2 circuit.

Check voltage at pin 3 cavity of the connector. You should measure battery voltage within one volt. No voltage here would indicate an open 5 ohm ballast resistor or bad connection. Refer to Figure 10-6C while making an ohmmeter check.

Next, check pins 4 and 5 cavities of the connector with an ohmmeter. You should measure pickup coil resistance between 350 and 550 ohms. Also measure the resistance of either pin 4 or 5 to ground. The resistance should be infinite. If the wiring and connections check out okay but the resistance measurements are wrong, the pickup coil should be replaced. When replacing the pickup coil, set the air gap between a reluctor tooth and the pickup coil core for .008″. Use a nonmagnetic feeler gauge so that the magnetism will not give you a false indication.

Pin 5 of the control unit should measure zero ohms to ground. Loosen and retighten the control unit hold-down screws if zero ohms is not measured. If continuity still does not show zero ohms, the control unit is defective.

Turn the ignition switch to ''off'' and reinsert the control unit connector. Turn the ignition switch on. Again check the spark, 3/16″, from the high-voltage coil secondary to ground while cranking the engine. If you still don't get a spark, replace the control unit. Check the spark again. If the spark is still missing the only thing left is the coil. Replace the coil and the problem should be solved.

When you're working on this type of electronic ignition system, remember that it resembles the conventional ignition in many ways. A systematic troubleshooting approach as outlined previously will make servicing this system as easy as the standard ignition. As in all electrical-electronic repairs, once you understand a system and use some common sense coupled with simple troubleshooting techniques, you will be successful.

11

REPAIRS FOR THE AUTOMOBILE STARTING SYSTEM

This chapter is about the electrical system used with automobile starters. Have you ever wondered just what happens to enable the car to
start? When the car is starting like a dream, the last thing in your mind
is starter operation. However, when the car doesn't start, you probably
would like to know a lot about the starting system . . . fast!

In this chapter you will read about the starting system and the
electrical controls that make it all happen. When you know how things
are supposed to work, it's a lot easier to fix them when they don't
work. The heart of the starting system is the storage battery. Here
you'll learn some of the things you can do to check and maintain the
battery. There's a section on the ignition switch dealing with its part in
starter operation. Relays and solenoids are also important elements in
the starting system. It's easy to understand electrical starter control
when you know about these devices. At the end of the chapter starting
motors are discussed. You may be surprised at some of the ways you
can repair starting motor troubles.

11-1. Tips on Proper Operation

Figure 11-1 shows a typical automobile starting circuit. The heavy current path for the starter motor is shown in wider lines. This particular starting system uses a relay and a solenoid to activate the starting motor. Here's the chain of events that takes place when the ignition switch is placed in the start position. As long as the transmission is in the neutral or park position the safety switch is closed. Electricity will flow out of the negative side of the battery to ground, then from ground through the closed safety switch through the relay coil to the momentarily closed starter switch and back to the positive side of the battery. This is a complete circuit that creates a magnetic field in the relay coil. The magnetic field will cause the N.O. (normally open) relay contacts to close.

As soon as the relay contacts close, a new circuit is developed. Now current will find another path from the negative side of the battery to ground, from ground to the solenoid coil through the closed relay contacts and back to the positive battery terminal. A magnetic field is created in the solenoid coil. This will cause the heavy solenoid N.O. contacts to close. Now the third and last starting circuit is completed.

Heavy current will flow out of the negative battery terminal to ground. From ground it will find a path through the starting motor and the solenoid contacts back to the positive terminal of the battery. The current passing through the starting motor will cause it to rotate and at the same time engage into the gears of the flywheel. This action will cause the flywheel to turn, operating the main engine. As soon as

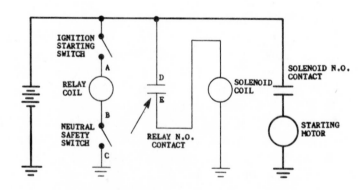

Figure 11-1. Typical automobile starting circuit.

ignition occurs in the engine (see section 10-1), the ignition starting switch is released to the N.O. position. Then the magnetic field around the relay coil will collapse, opening its contacts. This in turn causes the solenoid coil to lose its magnetic field and the solenoid contacts open removing power from the starting motor. The motor stops turning and a spring disengages it from the flywheel. The main engine is now running on internal combustion power.

Most cars have a similar starting arrangement. Some starting circuits eliminate the starting relay using the starting switch to directly control the solenoid coil. A solenoid or relay-controlled solenoid is always used to activate the heavy starter current. In cold weather the starter motor may need 200 or more amperes to operate. If the starting switch controlled this current directly it would have to have massive contact surfaces. Also, the large battery cable would have to enter the passenger compartment terminating on the starting switch. Then an equally large cable would be required from the other side of the starting switch back to the starter. This type of direct-control system would be very costly and bulky. Relays and solenoids are the answer; they do a fine job.

11-2. Checking the Battery

One of the best ways to check the automobile storage battery is with a hydrometer. This device checks the weight of the solution in the battery. Battery liquid is composed of water and sulfuric acid. A discharged battery loses most of the acid content leaving mostly a water solution. A fully charged battery has a high content of acid making the liquid heavier than water. The hydrometer body is calibrated in specific gravity readings (the weight of the solution compared to the weight of pure water). To measure the specific gravity some of the battery fluid is drawn into the hydrometer. Inside a calibrated float will indicate the condition of the liquid. Figure 11-2A shows a battery being checked using a hydrometer.

Most hydrometer floats are calibrated to indicate correctly only at one fixed temperature, 80°F. The battery acid will measure heavier when the temperature is lower and lighter when higher. Figure 11-2B shows a correction chart for reading a hydrometer at various temperatures. Here's a couple of examples of hydrometer readings at temperatures above and below 80°F.

Hydrometer reading	1.250
Acid temperature	90°F.
Add specific gravity	.004
Corrected reading	1.254

Hydrometer reading	1.250
Acid temperature	30°F.
Subtract specific gravity	.020
Corrected reading	1.230

A new, fully charged battery should measure around 1.280 specific gravity. Most hydrometers are calibrated from 1.160 to 1.320. Once you find the weight of the battery acid you can evaluate the

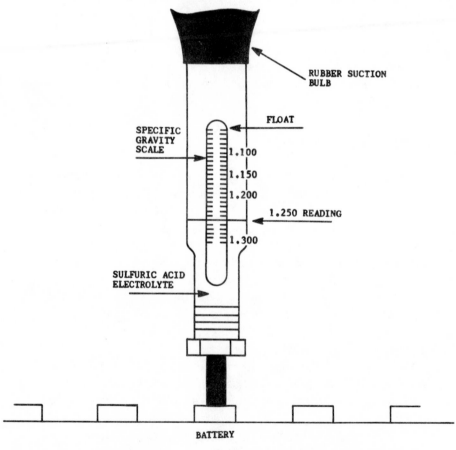

RUBBER SUCTION BULB

FLOAT

SPECIFIC GRAVITY SCALE

1.100

1.150

1.200

1.250 READING

1.300

SULFURIC ACID ELECTROLYTE

BATTERY

Figure 11-2A. Checking a battery using a hydrometer.

battery condition. The following list shows the relative condition of the battery compared with specific gravity of the solution.

Specific gravity reading	Battery condition
1.280	Fully charged
1.250	Good
1.225	Fair
1.150	Bad

Each battery cell should read approximately the same. A cell that reads .05 specific gravity below the other cells probably indicates a bad cell. A battery charge should increase the solution reading to at least 1.225. If you can't, the best bet is to replace the battery.

Check the battery solution after it is thoroughly mixed with any water that may have been added. Hold the hydrometer at eye level in a vertical position. Be sure the float is freely suspended in the solution not touching the sides or top and bottom. The top of the liquid will

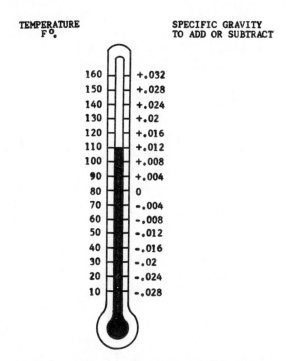

Figure 11-2B. Hydrometer correction chart.

match up with the specific gravity reading. Take the reading, make temperature corrections, and you'll have a good idea of your battery condition. Don't forget to check each cell in this manner.

A good thing to have around the house is an inexpensive battery charger. These chargers will give the battery a slow charge (the best method to charge a battery). They usually take 12 or more hours to do the job. After charging, the specific gravity should be noticeably better. You can continue slow charging a battery as long as the specific gravity reading keeps on increasing. The only danger in charging a battery is when the fluid temperature increases above 110°F. Most inexpensive battery chargers do not have enough current capacity to overheat the battery fluids. It's simple to hook up a battery charger; just connect the positive (red) lead to the positive battery terminal and the negative (black) lead to the negative battery terminal. The polarity of each battery terminal is generally stamped on the battery post or case. When working on automobile batteries, be sure to observe safety precautions (see section 1-2).

Take good care of a battery and it will generally give outstanding service. Always keep the fluid level above the plate area by periodically adding water. Generally ½" of battery fluid above the plates is sufficient. Most battery cells should be filled to the bottom of their filler holes. Distilled water is the very best thing to add to the battery. It's inexpensive and won't contaminate the acid with any impurities or leave any residue in the battery. Keep the top of the battery clean and free from any dirt and acid salts. A good way to clean a battery is to make a solution of warm water and baking soda. Scrub this solution over and around the top of the battery with a stiff brush. You'll see a lot of foaming action as the baking soda neutralizes the acid residue. Be sure you don't get any of this cleaning solution inside of the battery! Clean off with water and wipe dry. You won't believe it's the same battery.

Give the battery posts and cable clamps some attention. It's a good procedure to remove the cables and clean and burnish the posts and clamps. Take some rough sandpaper to shine up the posts like new. Wrap the sandpaper around a wooden dowel rod and clean out the inside of the cable clamps, too. When replacing the cable ends they should fit snugly around the posts. Tighten the cable clamps' screws reasonably tight. Don't excessively tighten the battery hold-down screws either or the battery case could crack; just tighten enough to keep the battery firm and stable in its cradle. After everything is cleaned and shines like new, apply some light mineral grease to the posts and clamps to help prevent future deterioration. A battery that is given this kind of attention once or twice a year will rarely let you down as long as the rest of the charging system is functioning normally.

11-3. Diagnosing Starting Switch and Relay Problems

The starting switch on most cars is part of the ignition switch assembly. Most ignition switches are key-activated with four separate positions: accessory, off, on, and start. When you turn the key to the start position and nothing happens, no sound, not even a click, one of the probable defects is the starting switch.

Figure 11-3 shows the circuit of a typical ignition switch. As soon as the ignition switch is placed in the start position, a circuit should be completed from the battery to the neutral safety switch, through the relay coil and start switch back to the battery. Anything that is defective in this series circuit would keep the start circuit inoperative.

Here's a simple way to check this circuit just using a clip lead. Make sure the gear shift lever is in the park or neutral position. Put the ignition switch in ignition #1 position. Fasten one end of the clip lead to the positive post of the battery (ungrounded terminal). Then with the other end of the clip lead momentarily touch terminal A on the starting relay. If the car starts normally, the ignition switch is defective in the

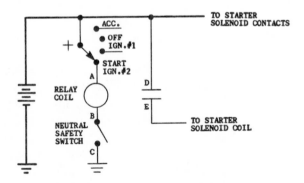

Figure 11-3. Ignition switch circuit.

start position or a lead or connection is defective. You can confirm a defective starting switch by crawling under the dashboard and jumping a clip lead from the + (hot) terminal of the ignition switch to the start terminal. The hot terminal of the ignition switch is the only terminal that measures + 12V in respect to the car frame when the switch is in the off position. The start terminal can usually be identified by wire color code. It will generally be the same color as the wire fastened to terminal A of the starting relay.

If the ignition switch checks out okay, try another clip lead test. Turn the ignition switch off. Fasten one end of the clip to ground and the other end to terminal B of the starter relay. The neutral safety switch may not be closing. The clip lead will bypass it. Now try to start the car again using the ignition switch. If normal starting is restored, you can bet the neutral safety switch is not closing. *Caution*: Don't forget to replace the switch because now the automobile will be able to start in any automatic transmission position.

Suppose none of the previous clip-lead tests restored starter operation. Maybe the relay coil has opened. You can simulate starter relay contacts closing by jumping across terminals D and E with a clip lead or screwdriver. This will energize the starter solenoid and activate the starter motor. If your car doesn't have a starter relay, jumper the two large terminals of the remote solenoid with a conductor capable of carrying the heavy starting current. Two large screwdrivers will generally work. Touch one screwdriver blade to one terminal and the other screwdriver blade to the other solenoid terminal. Then bring the screwdriver handles toward each other until the shafts touch. This will complete the starter circuit and operate the motor.

11-4. Testing the Starter Solenoid

Most cars have a solenoid mounted on or in the starting motor to engage the starter gear into the flywheel and to furnish the electrical path for the heavy starting current through the motor. Many solenoid troubles are caused by oil leaks that allow seepage onto the solenoid contacts. This will cause the contacts to burn and pit prematurely. Many solenoids can be rebuilt or repaired without too much trouble. Often solenoid contacts can be reversed so the unused side is available. Sometimes just rearranging the position of the contacts will cure the problem.

Suppose the starter solenoid is not energizing and you have already checked out the ignition starting switch circuit (see section 11-3). Jump a clip lead from the ungrounded battery terminal to the solenoid coil terminal F, as shown in Figure 11-4. The large terminal G on the solenoid is already connected to the battery with a heavy cable. Simply clip one end of the lead on this terminal and touch the other end to small terminal F. This will bypass the starting relay contacts and connecting wire. If the solenoid activates the starter motor normally, the solenoid and starter are okay. Some starter solenoids have a third terminal (small) H, which is used to provide unballasted ignition current during start. When you don't know which small terminal, H or F, is the coil connection, try momentarily touching the clip lead to either one. The one that activates the starter is terminal F. If neither terminal works you'll have to do some further troubleshooting.

Check out the solenoid coil if you don't hear the solenoid click when you make the above test. Measure the resistance between terminal F and the car frame. A high resistance or infinity will indicate an open solenoid coil. The resistance of a good solenoid coil should measure about .5 ohms. If the coil checks out bad, replace it with an identical unit.

If you hear the solenoid click when energized but the starter motor doesn't respond, chances are the contacts or the starting motor is defective. If the solenoid you are working on has two large terminals, short them together. This will bypass the internal contacts activating the starting motor. If the motor starts, the contacts are bad. If it doesn't, the starting motor is the problem. Solenoids that have only

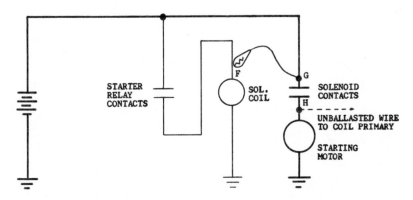

Figure 11-4. Typical starter motor solenoid circuit.

one large terminal on their case will have to be disassembled to deter-mine if the trouble is contacts or the starting motor.

11-5. Correcting Starter Motor Troubles

Before blaming the starter for failing, be sure to check out the battery with a hydrometer (see section 11-2). Once you are positive the battery is good, you can then check out the starter motor. A common trouble is poor battery cable connections. Turn on the lights and crank the engine. If the lights go out and you don't have any starting motor action, you probably have a bad connection. Make a DC voltage check with a VOM of the connections and cables (Figure 11-5A). Connect voltmeter A between the battery negative (grounded) post and the engine block. Crank the engine while monitoring the voltage. The voltage should be close to zero, not more than .2V. If higher voltage is indicated, you have a bad connection between the grounded battery post and the engine. Refer to Figure 11-5B and measure the voltage drop, voltmeter X, from the negative battery post to the clamp around it. Do the same between the terminal that fastens to the engine block, voltmeter Y. Also measure the voltage drop along the ground cable,

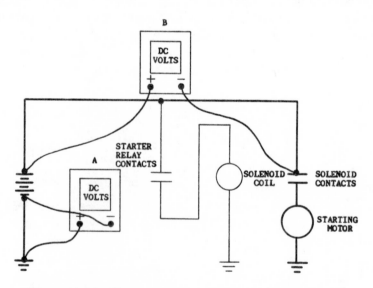

Figure 11-5A. Typical starting motor and solenoid circuit.

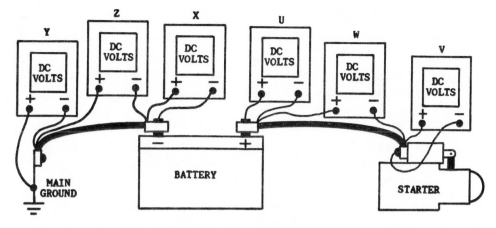

Figure 11-5B. Checking voltage for high-resistance connections.

voltmeter Z. Make each of these measurements while cranking the engine. The voltmeter that reads more than .2V will indicate the trouble area.

When the ground circuit checks out okay, make the same kind of voltage drop tests on the positive battery terminal, cables, connections, and starter terminal. Voltmeter B, Figure 11-5A, should not read more than .2V while cranking the engine. If more voltage is measured, refer to Figure 11-5B and make voltmeter checks U, V, and W. The meter that measures voltage indicates the bad connection.

When the starter operates sluggishly and the lights dim excessively, there is probably something causing the starter to work extra hard. A new or rebuilt engine could cause this symptom if the clearances were too close. Heavy crankcase oil and cold weather would also make starting difficult. Even poor timing or a defective spark advance mechanism could cause the cylinder to fire prematurely causing the starter to work against the engine. If any fluids leak into an engine cylinder such as coolant or fuel, the engine will be very difficult to crank. Since fluids do not compress, one cylinder loaded with fluid would stop cranking completely. If you suspect this form of trouble, simply remove the spark plugs and crank the engine. The engine will crank and fluid will spurt out of the leaking cylinder.

Many starter troubles are obvious. The following few things can happen:

1. Starter turns but does not engage flywheel.
2. Starter solenoid vibrates back and forth.
3. Starter operates but will not disengage.

Generally these faults are caused by mechanical breakage in the starter mechanism and are readily apparent if the starter is disassembled.

Remember, the starting system of an automobile is one of the most reliable parts of the car. Before you pull out that starter don't forget to check the battery, cables, and controls as suggested in the previous sections. Chances are you'll save yourself a lot of work.

12

SIMPLIFIED SERVICING
FOR ELECTRICAL DEFECTS
ON GAS FURNACES

Many gas furnace defects are electrical in nature and easy to fix. This chapter will explain how a typical furnace is supposed to work in a step-by-step fashion. Once the sequence of operation is understood, you will usually find the repair quite simple.

The brain of the heating system is the thermostat. In this chapter you will read about how this little device can direct furnace operation flawlessly for years. You'll learn how to troubleshoot and repair them when they become defective. Furnaces operate with automatic controls. You'll find these controls logical to understand and a breeze to repair. Limit switches, blower controls, gas valve solenoids are all automatic devices that can be checked easily.

Here you'll find out some of the problems that happen in the blower motor. On some cold, wintry night you'll be glad to understand furnace operation. Chances are you'll get it going again in a jiffy and save yourself some cold feet and a bundle of money.

12-1. How It Works

Summer is drawing to a close. Before going to bed you set the thermostat for 68°F. Sometime during the night the house temperature will drop below 68°F and the furnace will automatically start to heat. The air will be warmed and blown throughout the house and for the remainder of the night the furnace will maintain a 68°F temperature. It's a beautiful system when it is working.

Let's follow the previous heating sequence step-by-step to find out exactly what happens. Figure 12-1 shows a typical gas furnace schematic and pictorial drawing. Refer to this figure for the following explanation. Once the master on-off switch is turned on, AC current will flow through the primary of the step-down transformer, creating a magnetic field around the windings. The thermostat is set at 68°F, but the air temperature is 75°F so no other electrical action will occur. The manual gas valve is open so gas will be able to flow in the pilot tube. The pilot light is positioned very close to the main burner so it will

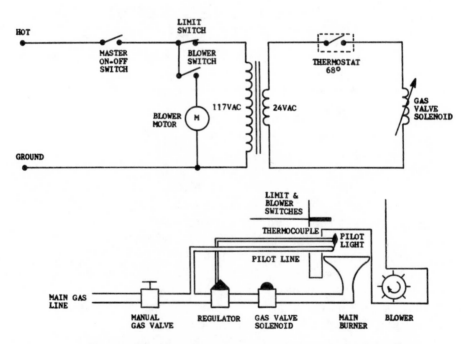

Figure 12-1.   Gas furnace schematic and pictorial drawing.

ignite the gas when the burner turns on. The pilot light is also heating the thermocouple, which is placed directly in the flame. As long as the thermocouple is heated, a small current will be generated, which allows the gas regulator to function. This is a safety precaution so that if the pilot flame goes out the main gas line is closed by the regulator.

When the temperature drops to 68°F, the thermostat, which is really nothing more than a temperature-activated switch, closes. This will enable AC current to flow in the 24V secondary circuit of the transformer. The gas valve solenoid will energize, allowing gas to flow into the main burner. The pilot light will ignite the main burner gases causing the heating chambers of the furnace to become hot. As soon as the temperature inside of the furnace reaches a preset amount, another automatic heat-activated switch, the blower control, will turn on the blower motor. The motor is usually connected to the blower assembly by a drive belt. As the blower turns, it will blow filtered air across the outside surface of the combustion chamber picking up the heat. The heated air will continue through the ductwork in the house and into the rooms. At the same time, a suction is created and room cold air will be returned to the furnace by the cold air ducts.

Shortly the thermostat will sense the increased air temperature and open. Immediately the gas valve solenoid will de-energize causing the automatic valve to close. The main burner will stop. However, the blower will continue to blow heated air through the system until the blower control switch cools down sufficiently to open. At that time the entire control system will return to the original standby operation. Nothing more will happen until the room temperature drops below the thermostat setting. Then the whole process will repeat.

In Figure 12-1 note the other automatic switch shown called the limit switch. This is a safety switch located in the top of the furnace plenum to limit the heated air temperature. The limit switch is heat-activated and preset to open at some temperature, usually around 200°F. From the furnace schematic you can see that when the limit switch opens, the blower will continue to force air through the system lowering the temperature. However, the limit switch shuts down the automatic gas valve so the air stops being heated. As soon as the air temperature drops, the limit switch will close allowing the air to be heated again.

Let's go back and talk a little more about the thermocouple. Most gas regulators are constructed so they will not pass gas unless the pilot flame is on. A thermocouple is made of two different kinds of metal joined together at one end. When the junction is heated a small amount of electricity will flow from one metal to the other. Inside the gas regulator a small electromagnet is powered by the thermocouple current. As long as current is flowing, the gas can flow freely through the regulator, but as soon as the thermocouple current stops, the electromagnet will release a regulator spring-loaded valve stopping gas flow. This is a very necessary safety feature to insure gas shut-down when the pilot is not lit.

12-2. Remedy for Thermostat Problems

The thermostat is simply a temperature-activated switch and a thermometer. The thermometer pointer indicates room temperature while another pointer shows temperature setting. A properly working thermostat will keep the room temperature at the same point as the temperature setting.

Most thermostats use a bimetallic spring and a mercury switch to control the gas valve (Figure 12-2A). A bimetal strip or spiral consists of two different kinds of metals joined together. When the spiral is heated, one of the metals expands more than the other causing the spring to move. This movement causes the mercury inside of the glass vial to break contact turning off the gas valve. As the spiral cools down, the mercury shifts position again turning on the gas valve. This

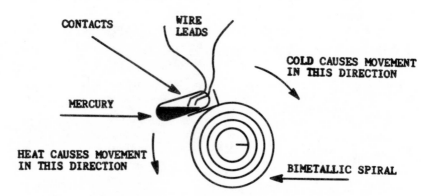

Figure 12-2A. Thermostat bimetallic spring and mercury switch.

process continues over and over to maintain a constant temperature. Most thermostats also contain an adjustable heater that provides the best timing for on-off cycles. Inside the thermostat you'll see a dial marked in amperes, usually .4 to .8. For best results this setting should match the current requirements of the gas valve. For instance, if the gas valve needs .5A to operate, the heater in the thermostat should also be set at the .5 position.

The thermometer portion of the thermostat is simply another bimetallic spiral with a pointer connected to its outside end. As the spring expands and contracts with temperature variations, the pointer will move along a dial indicating room temperature (Figure 12-2B). When the house temperature seems warmer or colder than the thermostat thermometer pointer indicates, the thermometer is probably reading incorrectly. Check it by placing an accurate room thermometer next to the thermostat. You may have to hold it in place with masking tape. Wait about ten minutes and then compare the two thermometers.

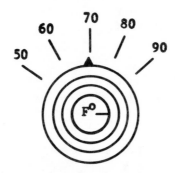

BIMETALLIC SPRING

Figure 12-2B. Thermostat thermometer.

If the thermostat thermometer is defective, its reading will vary from the other thermometer.

Some supposed thermostat problems have nothing to do with the thermostat or associated circuit. The location of the thermostat is very important to proper operation. A good location would fit the following list:

1. Mounted about five feet from the floor
2. On an inside wall near the center of the house

3. Good natural air circulation
4. Away from hot spots
   a. warm air ducts
   b. fireplace
   c. lamps
   d. television
   e. sun rays
5. Away from cold spots
   a. outside walls
   b. drafts
   c. unheated room on other side of partition
   d. unheated air in the wall
6. Dead spots
   a. behind doors
   b. corners

Suppose the thermostat does not turn on the gas valve at any position of the temperature dial. Remove the thermostat cover and short across the two connection terminals with a clip lead. If the gas valve solenoid clicks on and the main burner ignites, the thermostat is defective. If still nothing happens, the thermostat is probably okay and some other part of the circuit is defective.

Figure 12-2C shows a partial circuit being controlled by the thermostat and how to check it. Check the primary AC voltage of the step-down transformer with a VOM (VOM A, Figure 12-2C). It should measure 117 VAC. Next, check the secondary, VOM B, for 24 VAC. Feel the transformer case. It should be warm but not hot. If you measure low secondary voltage and the transformer case is very hot, something could be loading it down. Remove one of the 24V secondary wires from the transformer. If the secondary voltage returns to 24 VAC and the transformer case cools, the transformer is okay and something else is defective.

If the transformer checks out good, measure the voltage across the gas valve solenoid coil, VOM C. The meter will measure 24 VAC when the thermostat is on and 0 VAC when it's off. You should hear the click of the solenoid as the gas valve activates. The solenoid coil is probably open when 24 VAC is present but there is no solenoid action. Measure the coil with an ohmmeter with power off to check continuity. A good solenoid coil will measure around 10 to 20 ohms. A coil that measures high or infinite resistance is probably open. Often the wire

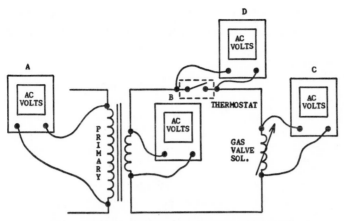

Figure 12-2C. Thermostat-controlled circuit.

will break at the point where it exits from the coil housing due to the solenoid vibration over the years. This type of break can be easily fixed by removing the coil cover and splicing the connecting wire back again. Provide some kind of strain relief with electrical tape or string so it won't happen again.

VOM D, Figure 12-2C, should measure 24 VAC when the thermostat is off and 0 VAC when the thermostat is on. When the thermostat, transformer, and gas valve solenoid check out okay, the problem is probably in the connecting wires. These wires seldom break or short but it can happen. The most likely place to check for breaks is close to where the connections are made. These are the places where there is vibration and the wire could have been weakened by a nick at installation time. A continuity check will test the wires in the walls.

12-3. Checking the Thermocouple

Whenever the tip of the thermocouple is in the pilot flame, the regulator will allow gas to flow into the main burner. The small current generated in the thermocouple will electromagnetically keep a safety valve open in the regulator. If the pilot flame goes out or is blown away from the thermocouple tip, the regulator safety valve will close and prevent unburned gas from entering the combustion chamber.

A thermocouple that becomes defective will lose its ability to generate current from heat. Even when the thermocouple tip is directly in the pilot flame, the regulator will shut off gas flow. Many regulators have a manual method to open the spring-loaded safety valve simulating thermocouple action. Of course, the valve will automatically close when manual force is released.

If the thermostat is controlling the gas valve normally (you'll be able to hear and feel the gas valve solenoid actuate with the setting of the thermostat) and no gas is entering the combustion chamber, the trouble is probably a defective thermocouple. Make certain the pilot light has not just blown out and that the thermocouple tip is positioned directly in the pilot flame.

The thermocouple is usually quite easy to remove. Generally it is held in a bracket in the combustion chamber with a retaining clip. The other end of the thermocouple is screwed into the regulator with a small pipe fitting. Pull off the retaining clip and unscrew the fitting. Slip the thermocouple from its bracket and with a little maneuvering it will be free.

You can make a couple of checks on the thermocouple to test it. A VOM set on R × 1 resistance range should measure zero ohms between the copper outside tube and the inside conductor at the regulator end of the thermocouple. This will not detect a short but will show if the thermocouple is open. Figure 12-3 shows how the thermocouple can be tested using a VOM and a torch. Clip one VOM probe to the copper tube and the other probe to the inside conductor. Set the VOM for DC volts at the lowest voltage range. Heat the tip of the thermocouple with the torch until it is red hot. You should measure between .05 and .1 DC volts. Reverse the VOM probes if the meter reads

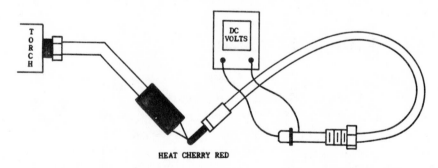

Figure 12-3. Testing a thermocouple using a VOM and propane torch.

backwards. If you don't measure this voltage, the thermocouple is definitely bad.

Most hardware stores keep replacement thermocouples in stock since they are relatively standard items. Just be sure to get one that is as long or slightly longer than the original. You can always shorten the thermocouple with a coil or two, but you can't lengthen it.

Install the replacement thermocouple and your furnace will probably work like new again. If the regulator still shuts off the gas flow, the regulator itself will have to be replaced.

12-4. Testing Blower and Limit Controls

The blower and limit controls are generally actuated by a bimetallic helix mounted inside of the furnace plenum. Figure 12-4A shows a typical blower and limit control assembly. The controls can be adjusted for a range of temperatures indicated on the dial. The blower switch generally is turned on at about 125°F. and off at 100°F. The limit switch turns off at about 200°F. They are adjustable so optimum settings can be made for various applications.

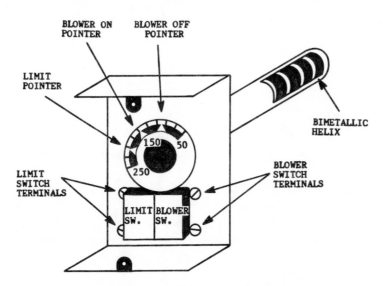

Figure 12-4A. Blower and limit control assembly.

Figure 12-4B shows the schematic for one type of blower and limit switch circuit. The limit switch is a normally closed SPST and the blower switch is a normally open SPST. When the bimetal helix reaches a set temperature, the blower switch is turned on automatically so the hot air is moved throughout the house by the fan. It will turn off at a somewhat cooler temperature. The limit switch is also actuated by the bimetal helix, opening up when the furnace plenum gets too hot. Opening the limit switch removes power from the transformer shutting down the gas valve and stopping the main burner combustion.

When the furnace is heating but the blower fan doesn't start, one of the troubles could be the blower switch or the actuating helix. Jumping a clip lead across the blower switch terminals will start the blower motor if this is the case. If the motor starts then you have to determine if the trouble is the switch or the helix. Generally you can reach the switch button with a small screwdriver to actuate the switch manually. Proper blower operation with manual switch action indicates a bad bimetallic helix or misadjustment. Of course, if the switch won't turn on the blower motor when pressed manually, the blower switch is

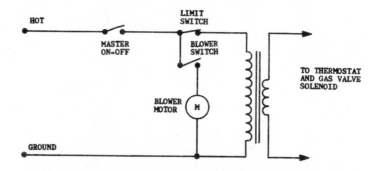

Figure 12-4B.   Blower and limit switch circuit.

the culprit. If none of the previous tests operate the blower motor, it is probably defective (see section 12-5).

You can check limit switch operation by connecting the VOM set at 117 VAC across the transformer primary. Voltage should be there until the limit switch is manually actuated and then drops to zero. This will check the switch. The bimetal helix can be checked in a couple of ways. You can disconnect the blower motor so the furnace will heat hot enough to reach the limit setting of around 200°F. At this time the voltage across the transformer primary will drop to zero and the main burner will shut down. Another method to check the limit switch action is with a heat lamp. Pull the switch and helix assembly from the furnace and position it so you can shine a heat lamp on the bimetal strip. The limit switch will open automatically when the limit temperature is reached. Don't forget to observe all electrical safety rules while making these checks since 110 VAC is generally exposed at the switch terminals.

12-5. Working with the Blower Motor

Most blowers are powered by a separate motor with a belt drive. The two most common problems related to the blower and motor are a dirty filter and a defective drive belt. Both of these problems are obvious after a visual inspection. A dirty filter will seriously impede air flow through the system resulting in inefficient furnace operation and inadequate heating. Belts can break or stretch. When checking a belt be sure to turn off power so the motor doesn't start while your hands are in the way. Look the belt over carefully for cracks or other signs of deterioration, especially on the inside area of the belt. You'll save yourself a future headache by automatically replacing the belt whenever deterioration begins to occur. Don't tighten the belt excessively when replacing it. You should be able to push in the belt about one inch at the middle when proper tension is made (Figure 12-5A). Too little tension will cause slippage and excessive wear. Too much tension could overload the motor wearing out the bearings and producing noise.

Most motors and some blowers have oil cups on each end of the shaft. They should be oiled twice a season with a good grade #20 motor oil. Don't overoil since a few drops are generally all that's necessary. If you are having noise problems with the blower-motor unit, the first step would be to check the belt for condition and tension. If that is not the problem, switch off main power, remove the belt and spin the motor and blower by hand while listening for noise. Grab the pulleys, putting side pressure on them. There should be no movement. If there is side play, the pulley could be loose (held in place with an Allen head screw) or the bearings could be worn.

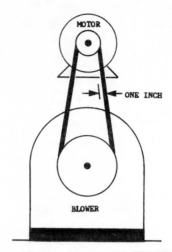

Figure 12-5A. Checking belt tension.

Blower motors are generally split-phase or capacitor start motors. Refer to sections 6-5 and 7-5 for methods to troubleshoot and fix these kinds of motors. If the motor is not working or buzzing or indicating any sign of life, don't automatically assume the motor is bad. Maybe the motor is not getting the power. Check the voltage at the motor terminals with a VOM. Figure 12-5B shows a typical blower motor circuit. If VOM A reads 110 VAC and the motor is not working, the

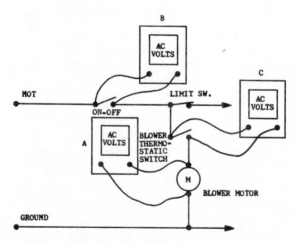

Figure 12-5B. VOM voltage checks for the blower motor circuit.

motor is probably defective. To be certain, remove the motor from the circuit and apply 110 VAC directly to the motor terminals with an AC clip lead (see section 2-1). If the motor still doesn't run, it is absolutely defective. However, if VOM A doesn't read any voltage, the master on-off switch or the blower switch could be the problem. VOM B should read zero volts when the switch is closed. VOM C should also read zero volts when the switch is closed. Don't forget that VOM C should read 110 VAC until the furnace plenum heats enough to activate the blower switch (see section 12-4). If either the master on-off or the blower motor switch is not closing, simply jump them with a clip lead to confirm your suspicions. If normal operations are restored, it will be a fairly easy job to replace the defective switch.

13

EASY-TO-MAKE
TELEVISION REPAIRS

There are many simple things that you can do to restore television operation. Reading this chapter will not make you an expert television repairman, but it will enable you to better understand television operation and how to fix a multitude of common TV problems.

You will learn how to diagnose and localize trouble. Fixing anything can be much simpler when you know exactly where the trouble originates. Since the television receiver is so complicated, proper trouble diagnosis and localization is essential for a quick repair. Don't get that hopeless feeling when the TV doesn't go on when you want to watch a favorite program. You'll be surprised how many common-sense repairs can be made to restore operation.

TV controls make super troubleshooting aids. Control operation can often lead you directly to the problem area. Here you will find the secret to using controls for troubleshooting tools. If you have ever checked all the tubes in a television receiver, you know there must be a better way. There is. In this chapter you will read about tube checking techniques that can really simplify this tedious task. One of the parts of a TV that requires the most service is that section behind the channel selection knob, the tuner. Most tuners are electromechanical devices that are prone to dirt and wear and tear. You'll be able to repair most tuner problems in a jiffy. One of the saddest words a TV set owner hears is when the repairman says, "You need a new picture tube." Read about some of the things that you can do to prolong the life of a picture tube. And who knows? If you can't repair it, why not replace it yourself? It is easier than you might think.

13-1. Understanding What Makes a TV Tick

Plug a TV into a wall outlet; turn it on and presto, in short order you're watching a moving picture, probably in color, with sound. It's magic! Well, maybe it's not magic, but it certainly is amazing that this happening can occur.

In order for a television receiver to operate, it must receive two kinds of electrical energy: electricity for power and electromagnetic waves for signal. Power is furnished by wall outlets, AC, or by a battery, DC. In the TV the power is used to operate the multitude of stages necessary to produce picture and sound. The signal energy is created at the television broadcasting studio and represents all the ingredients needed for picture and sound. Some form of antenna is necessary to capture part of this radiated signal for the TV receiver. Once the signal is reaching the receiver, it is amplified and conditioned in the various stages to produce sound, picture, and color.

Let's talk about the electricity for power. DC power is needed inside of the television so all receivers have one power supply section that changes AC electricity to DC. Another power supply is also needed to produce very high DC voltage for the picture tube. All picture tubes need an extremely high DC voltage in order to produce brightness. There's another power section that creates the energy to produce the vertical deflection portion of the brightness. Also, a power section is needed for the horizontal deflection portion of the brightness. A term used to describe the brightness of the picture tube is called the "raster." The raster is the lit face of the picture tube when no video signal is present.

A television receiver that is receiving power but no signal will have a raster but no picture (video). There will be no sound but you should hear some noise from the speaker when the volume control is set at maximum. Here's a list of the power sections of a typical television receiver:

Section	Purpose
Low-voltage power supply	Changes AC to DC to power all receiver stages
High-voltage power supply	Builds up the low voltage to high voltage to light the raster
Vertical deflection	Creates the energy needed to fill the raster vertically
Horizontal deflection	Creates the energy needed to fill the raster horizontally

The signal coming in the antenna furnishes information needed for the picture, sound, synchronization, and color. The picture information is called the video signal. The sound portion of the signal eventually results in the audio signal, which comes from the speaker. The synchronization signal is necessary to synchronize the transmitter with all the receivers for perfect harmony in picture display. The color information controls the color intensity, hue, and synchronization. When all the power circuits are functioning with all the signal circuits, the end result is modern television.

The following list shows the main signal section and the purpose:

Section	Purpose
Video chain	Amplifies and conditions the video signal to produce the picture
Sound chain	Amplifies and conditions the sound signal to produce the audio
Synchronizing	Synchronizes the vertical and horizontal sections in step with the transmitter
Color	Amplifies and conditions the color signal to produce the color portion of the video

The picture is put on the picture tube spot-by-spot in proportion to the video signal by the vertical and horizontal deflection circuits, which are in turn controlled by the synchronization signals. The face of the picture tube is coated with phosphor particles that glow when struck by the electron beam. In color TV the phosphors produce colored light. When the electron beam strikes the phosphor hard it will glow with much intensity, when struck with less energy the phosphor will glow with medium intensity, and when not struck it will not glow at all. The combinations of all the phosphor particles glowing in proportion to the strength of the video and color signals at the same time produce the picture.

The electron beam in the picture tube that writes all picture information is called the scanning beam. The scanning beam moves across the picture tube from left to right much like the words on this page. The horizontal deflection circuits control this movement as well as the return of the beam to begin writing the next line. The vertical deflection circuits move the beam down slightly as the line is written so each new line is slightly under the previous one, again just like the printed page. When the last line is written at the bottom of the picture tube, the vertical section moves the beam back up to the start and the next picture is begun.

You cannot see the picture being made dot-by-dot or line-by-line because of the extremely high speed of the electron beam and the fact that the phosphors continue to glow long after the beam has left. The image on your eye also has the tendency to remain even after the original picture has passed. This phenomenon is called persistence of vision. These factors make it possible to see a complete, moving color picture with no hint that you are just seeing millions of dots all glowing at different intensities and colors.

13-2. How to Diagnose and Localize Trouble

You're watching your favorite TV program, just becoming interesting, and all of a sudden the picture dies. What can you do to restore proper operation? A TV set is composed of many sections or blocks each having a definite part to play in overall operation. The secret to repairing a TV receiver is to learn the functions of the various blocks and how they work together to produce the sound, picture, and raster.

Figure 13-2 shows a block diagram for a typical color television

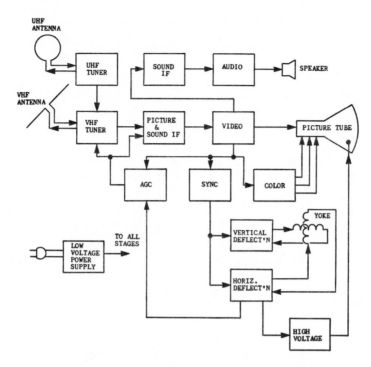

Figure 13-2. Color television block diagram.

receiver. This is a simplified block diagram but it shows all the important subsections. Let's examine each major TV block and describe its main trouble symptoms as related to sound, picture, and raster. Note that each block has one or more specific functions for receiver operation.

Block	Function	Trouble Symptoms		
		Sound	Picture	Raster
Low voltage power supply	Supplies power to all stages	None Poor	None	None
High voltage power supply	Supplies power to light picture tube face	Normal	None	None
UHF tuner	Selects UHF channels	No UHF (VHF normal)	No UHF (VHF normal)	Normal

Block	Function	Trouble Symptoms		
		Sound	Picture	Raster
VHF tuner	Selects VHF channels	None Weak	None Snowy	Normal
Sound section	Detects and amplifies sound	None Distorted Weak	Normal	Normal
Picture and sound	Amplifies picture and sound	None Distorted Weak	None Weak	Normal
Video	Detects and amplifies picture	Normal	None Weak	Normal
AGC automatic gain control	Keeps contrast and volume constant for all channels	None Normal	None Excessive contrast	Normal Normal
Sync	Synchronizes vertical and horizontal deflection	Normal	Rolling and tearing	Normal
Color	Supplies color to picture tube	Normal	Abnormal color No color	Normal
Vertical deflection	Supplies energy for vertical deflection of raster	Normal	Normal (If raster was present)	White horizontal line Margin at top and bottom of screen
Horizontal deflection	Supplies energy for horizontal deflection of raster	Normal	Normal (If raster was present)	None Margin at each side of screen
UHF antenna	Picks up UHF stations	None Weak	None Weak Snowy	Normal

Block	Function	Trouble Symptoms		
		Sound	Picture	Raster
VHF antenna	Picks up VHF stations	None Weak	None Weak Snowy	Normal
Speaker	Provides sound	None Weak Crackly	Normal	Normal
Picture tube	Provides picture and raster	Normal	Weak None Washed out Poor contrast	None Weak

Always observe the sound, picture, and raster symptoms carefully when evaluating a TV trouble. Then try to relate these symptoms to the block diagram. In most cases you'll be successful in pinpointing the defective area.

All TV receivers have a tube or block diagram layout pasted somewhere on the inside of the cabinet. Once you have identified the guilty section or sections, it's a simple matter to check or substitute the tubes, transistors, or modules that are in that section.

For instance, suppose the raster collapses to a thin, horizontal white line but the sound remains normal. Checking the block diagram chart we find that the vertical deflection block would produce this kind of symptom. If tubes are used in the set you're fixing, one tube generally is all that's used for vertical deflection. After locating the vertical tube, replace it and the chances are operation will be normal again. Do the same for vertical transistors if they are in sockets. In many modern TV receivers the vertical section is one plug-in module. Simply pull out the vertical module and plug in a replacement. Most TV manufacturers have some kind of trade-in plan for these modules.

13-3. Making the AC Power Check

When a TV receiver is completely dead, no sound, picture, or raster, the first thing you should check is the AC power. Countless TV repairmen have been called out to a house simply to plug in the television line cord. Moving furniture, house cleaning, and kids playing are ways of removing the line cord without anyone's knowledge. If the line cord is plugged into the outlet, be sure the outlet has power. Maybe the fuse or circuit breaker in the entrance box has opened from an overload on another part of that circuit. Check the outlet with a test lamp or VOM. If the outlet is okay, look for a circuit breaker button on the back cover. Occasionally circuit breakers will pop from a momentary overload. If pushing the button restores normal operation, your problems are over. If the button pops out again, read on. Next, remove the back from the TV. You'll probably need a cheater cord to bypass the receiver's line cord, which is generally built into the back cover. Cheater cords are readily available at any local electronic store. A line cord with well-insulated alligator clips on one end will also do the job. If the TV comes back to life with the cheater cord, the original line cord or plug is the culprit.

Once the back is removed and the set is being powered by the cheater cord, observe any tubes in the receiver for filament power. If power is reaching the TV, the filaments will be lit. For solid-state sets check the picture tube filament, which can be observed in the neck of the tube near the base. In some all-tube receivers the filaments are wired in series so that if any filament burns out all tubes will be out. Sets that have a series filament string will generally have the filament circuit identified on the tube layout chart (see Figure 13-3A). Notice the filament pin numbers and wiring sequence. To check the filaments measure the pins with a VOM set on R × 1 range. Tube pins are always counted clockwise from the bottom starting at the first pin to the right of the tube keyway. When the filament is okay you'll read a low resistance. Check each tube including the picture tube for an open filament. In a set of this type a filament trouble is the most likely kind of trouble.

TV sets that have tubes wired in parallel do not all go out when one filament burns out. You can recognize a parallel filament hookup when all the tubes begin with the number 6 or 12. Also you'll see a large power transformer mounted on the chassis. It's easy to spot the dark tube when its filament opens up. If you are not sure whether the filament is heating, simply feel the glass envelope. The glass will be hot if the filament is normal. *Caution*: Be careful when touching some tubes since they run quite warm. Also be certain not to touch any

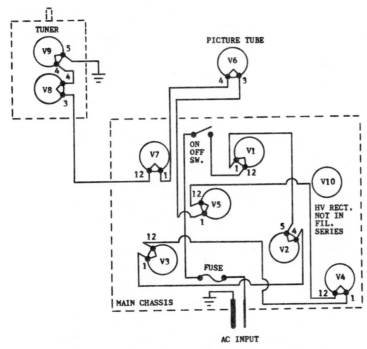

Figure 13-3A. TV series circuit tube layout chart.

high-voltage plate caps. If you look into the high-voltage cage and see the high voltage rectifier apparently not being heated, don't assume its filament is burned out. Most of these tubes will show no indication of filament light.

Suppose after you have removed the back cover and are operating the TV on the cheater cord, you don't see any filament glowing, not even in the picture tube. Since you've checked the outlet and are using a good cheater cord you know 110 VAC is present at the TV chassis. If the set had a circuit breaker you have already tried to reset that to no avail. Look around for a fuse or fuses. Pull them and check for continuity with the VOM on R × 1 range. Don't rely on a visual inspection of the fuse since it's common for the fuse element to open close to the end cap where it's impossible to see. Some TVs use a fusistor as a protective device. This is a combination fuse and resistor (Figure 13-3B). Most of these devices are located very close to where the AC comes into the TV set. They usually look like a flat or rectangular power resistor and are usually mounted with plug-in terminals. A good

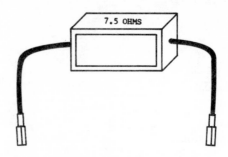

Figure 13-3B. Fusistor.

fusistor will generally measure under ten ohms. Replace it with an identical unit because their value is important for maximum protection.

Back to the circuit breaker problems. If a circuit breaker pops and you reset it and it pops almost immediately again, there probably is an overload in the TV. However, if the circuit breaker doesn't pop again for an hour or more after resetting, the circuit breaker itself may be defective. Most circuit breakers only require two connections and are usually easy to replace. Just be sure to replace it with one with an identical current rating. When in doubt about a circuit breaker, replace it. They are very inexpensive and you might save yourself a big repair bill.

13-4. Practical Ways for Checking Controls

The operator and service controls can often give valuable clues as to trouble areas. An inoperative or unusual-acting control will very often identify the defective section. The following list shows the block in which each control is located:

Control	*Section*
1. Fine tuning	Tuner
2. Contrast	Video
3. Brightness	Video or picture tube
4. AGC	AGC
5. Screens	Picture tube
6. Gain or drive	Picture tube
7. Focus	Picture tube, high voltage
8. Vertical hold	Vertical deflection

Control	Section
9. Height	Vertical deflection
10. Vertical linearity	Vertical deflection
11. Horizontal hold	Horizontal deflection
12. Width	Horizontal deflection, picture tube
13. High-voltage adjust	High voltage
14. Horizontal linearity	Damper
15. Volume	Sound
16. Tone	Sound
17. On-off	Low-voltage power supply
18. Circuit breaker	Low-voltage power supply
19. Color	Color
20. Tint	Color
21. Color killer	Color

Some of these controls are easy to adjust requiring no special equipment. Others should best be adjusted by a television technician. Here's how to adjust the common controls so you can evaluate the control performance.

Fine tuning	Rotate the control clockwise until the picture develops a slight herringbone pattern, then turn counterclockwise for best picture and sound.
Contrast	Rotate until the difference between blacks and whites are most pleasing to you.
Brightness	Rotate for overall picture tube illumination.
AGC	Tune in the strongest channel in your area. Rotate AGC control CW until the picture has excessive contrast and breaks apart. Then turn CCW for best picture.
Screens	Set the three red, green, and blue screen controls fully CCW. Move the Normal-Service switch to service position. The raster should collapse causing a dark screen. Adjust each screen control CW

	until a thin horizontal color line is visible in the center of the raster. When all screens are adjusted properly the line should be white.
Gain or drive	Adjust the two or three gain controls maximum CW that will produce a grey raster on an "off" channel.
Vertical hold	Slowly turn the control until the black horizontal bar is visible. Then rotate the control until the bar drifts upward and locks in place off the raster.
Height	Adjust the control until a black margin is visible at the bottom of the raster. Turn to just make the margin disappear.
Vertical linearity	Set the control until a black margin is visible at the top of the raster. Turn to just make the margin disappear. Notice that the height, vertical linearity, and vertical hold adjustment will slightly interact.
Horizontal hold	As you turn the control the picture should stay locked in horizontally for about a half rotation of the control. The entire raster may shift slightly horizontally. When the control is out of range the picture will break into sloping horizontal black bars. Reset the control so it is set in the middle of its hold-in range.
Width	Adjust control for horizontal raster size. Some sets have a sleeve on the neck of the picture tube for the width adjust. Slide the sleeve back and forth for the width adjustment.
Color	Clockwise rotation should increase color intensity.
Tint	Clockwise rotation should change the tint or hue of the colors. Adjust for best flesh tones.

Color killer	Turn control CCW. Set channel selector to an "off" channel. The "snow" or "confetti" should be color. Rotate the control CW until the raster is black and white.

Once you locate a control or controls that do not respond properly, check or substitute all plug-in tubes, transistors, or modules in that area. This common-sense method of using the controls to pinpoint defective areas works and can save a bundle of time that otherwise would have been spent checking good areas of the TV.

When you run across an erratic control that seems to work properly but introduces noise in the sound or flashes in the picture whenever it is adjusted, the control is probably dirty. A squirt with tuner spray can usually do wonders. Spray into the control through the opening near the connecting lugs. If the control is sealed or you want to clean it without removing the back cover, simply position the control so that gravity will allow the cleaning fluid to run down the shaft into the dirty area when the shaft is sprayed (see Figure 13-4). Work the control back and forth a few times and it will probably clean up like magic.

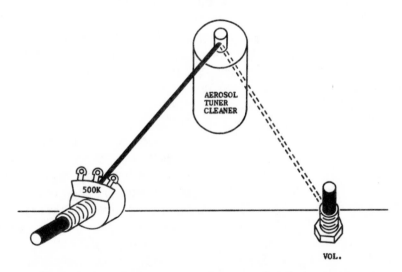

Figure 13-4. Cleaning dirty controls with tuner cleaner spray.

13-5. Simplifying Tube Testing

Tubes are the most unreliable part of a television receiver and are the most likely culprits when a tube-operated TV fails. Most troubles can be diagnosed to a particular section or sections (see section 13-2), so only a handful of tubes could be the source of the trouble. Generally it is a waste of time to check every single tube in a television set.

Some tube problems are readily apparent. In parallel filament circuits (section 13-3), the unlit or cold tube will be easy to spot. In a series filament string all tubes will go out when the filament opens. Section 13-3 explains how to find the bad series tube using the tube layout chart and ohmmeter.

It's fairly common for the glass to crack, especially the larger power tubes. The crack may not be apparent but the air leak will cause a chemical reaction inside of the tube, causing a whitish deposit to appear on the inside of the glass envelope. Any tube with this appearance must be replaced.

Some tubes will arc or flash in the center of their elements. This is definitely not normal and indicates that a new replacement tube is needed. In fact, if the arcing is allowed to continue there's a good chance some of the circuit will also be damaged. Another technique to easily locate bad tubes is tapping with an ordinary lead pencil while watching the tube, the picture, and listening to the sound. Tap lightly around the circumference of the tube. Any loose or intermittent connection will act up when tapped, indicating the problem as arcing in the tube or changes in the picture and/or sound. Picture tubes sometimes develop internal shorts between their electrodes. If this is the case, tapping around the neck of the tube near the base will sometimes dislodge the short fixing the tube.

When tube defects are not visually or physically apparent, they will have to be checked on a tube checker or replaced with new tubes. Most electronics stores and discount stores have tube checkers for the general public to use. A tube checker is not a reliable method for finding a faulty tube. They do not test tubes under actual operating conditions and only give a rough indication of the tube's quality. Many tubes that check bad on a tube checker will work properly in a circuit. Other tubes that check good will not work in a circuit.

Here are rough guidelines to follow so you can evaluate a tube's condition:

Type of tube	Tube checker's indication	Possible condition
Small tube but not an oscillator	Weak emission or gain	Probably okay
Small tube	Shorts on one or more elements	Replace
Small oscillator tube	Checks okay	May not work in circuit
Large tube	Weak emission or gain	Probably okay
Large high-voltage tube	Checks okay	May not work in circuit
Large tube	Shorts on one or more elements	Replace

When you're checking the tube be sure to tap the tube lightly while you're checking for shorts. A short indication is definitely a sign of a bad tube. A tube that checks low on emission or gain may or may not be usable. All tubes that operate in the high-voltage circuits such as the horizontal deflection amplifier, the damper, the regulator, and high voltage rectifier can only be reliably tube-tested when they have checked shorted or dead. If they check good you must substitute them with new ones to be sure. The only fully reliable test for any tube is a substitution with a known good tube. A known good tube is a tube that has already worked in a similar circuit. The majority of new tubes are good but occasionally you'll run across a bad one. It would be nice if you had a known good replacement for every tube in your TV receiver but in most cases this is not possible. Hopefully you can make a fairly accurate judgment of a tube's true quality by the kind of TV trouble symptom, the tube checker's indication, the type of tube, and a certain amount of pure luck.

Some troubles appear to be caused by tubes but are really not. For instance, if a trouble symptom appears when a tube is jarred or tapped, the real problem could be the connections between the tube pins and the socket. Sometimes the connections will become loose or oxidize. Often this defect can be repaired by simply moving the tube in and out a number of times to break the oxidation film. If this doesn't do the trick, scrape the tube pins with a knife and shoot some tuner cleaner into the tube socket.

Here's a good method to let the TV set check a tube. Often the same tube is used in a number of sections of a TV receiver. If you suspect this kind of tube to be defective simply interchange it with an identical tube in another section. If the trouble symptoms stay the same the tube is okay; if they change the tube is probably defective.

Make certain when replacing tubes back into their sockets that you have the right tube. It's easy to mix them up especially if you've pulled a number of tubes at the same time. Often it's a little difficult to insert the tube because of locating the keyway spacing. Most tube layout charts show where the keyway is located and really are helpful for those tubes buried deep inside the chassis. A flashlight and a small inspection mirror can come in handy, too.

13-6. Guide to Tuner Troubles

One of the most common types of troubles in a television receiver is that of a dirty tuner. Here's a list of trouble symptoms that could be caused by a dirty tuner:

1. Erratic channel selection
2. Picture flashes and noise in sound as side pressure is applied to the channel selector knob
3. Channel tunes in at a place other than the "click" position
4. Doesn't receive all channels

Most tuners can be cleaned quite easily with a can of aerosol tuner cleaner. Sometimes you won't even have to remove the back of the TV cabinet. Pull off the channel selector and fine-tuning knobs. As you look into the hole, you should be able to see the tuner chassis. With power off, squirt some tuner cleaner into the tuner crevices and holes that are accessible through the opening (Figure 13-6A). Then rotate the channel selector back and forth about twenty-five times. This should solve the problem. However, if the tuner problem remains, the spray probably has not reached all the contacts inside of the tuner. Remove the back from the TV and spray some more tuner cleaner into the sides and rear of the tuner chassis. Don't overspray since three or four bursts

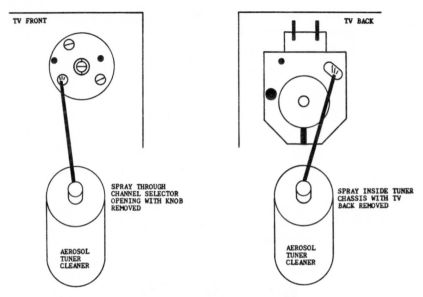

Figure 13-6A. Cleaning a tuner with aerosol tuner spray.

are usually sufficient. Again rotate the channel selector around and around to help clean the contact surfaces. Turn on the set for a beautiful picture.

If there is still evidence of bad contacts the only alternative left is to remove the tuner cover. It generally just snaps off. Turret-type tuners are the easiest to physically clean. If you have this type you'll see plastic strips mounted in a drum arrangement. The drum will rotate as the channel selector is turned. The contacts on the strips will be readily apparent. They should be shining silver. They will probably be oxidized and discolored black. To clean them squirt some tuner cleaner on a soft clean rag and briskly rub the contacts. They should clean up like new. Rotate the channel selector so you can clean all the contacts. If you run into a few contacts that just won't clean up, try rubbing them lightly with a typewriter eraser. This will usually take care of the most stubborn contacts.

Tuners with wafer switches cannot be cleaned as you would with a turret-type tuner. Try shooting aerosol spray directly at the contacts throughout the tuner assembly as you rotate the channel selector. This action will generally clean up this type of tuner.

Many tuner problems are not really tuner problems but stem from faults in the antenna system or lead-in wire. Antenna, lead-in wires, and tuners all present similar trouble indications when they are defective. Here are some symptoms:

1. Poor reception on all channels
2. Snow on some channels
3. Lack of color
4. Reception on one or more channels but not others
5. No picture, no sound, normal raster
6. Snowy raster, no picture or sound
7. Ghost in picture
8. Picture pulling

You can usually determine whether the problem is an antenna or tuner problem by substituting a rabbit-ears antenna for the existing one. If the trouble remains the same the tuner is at fault. A noticeable improvement indicates an antenna or lead-in wire malfunction.

If one or more channels will not fine tune properly anymore, the tuning screws or slugs for those channels probably need resetting. There are a number of ways to fine tune a tuner. Many of the problems result from a mechanical defect in the fine-tuning mechanism. Pull the tuner and inspect the fine-tuning arrangement carefully. If the problem

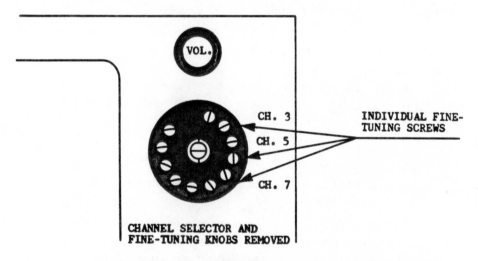

Figure 13-6B. Fine tuning adjustments accessible in channel selector opening.

is a mechanical one you'll probably spot it easily while you're looking at the "bad" channel.

Some TVs have fine-tuning slugs accessible through the opening left when the channel selector and fine-tuning knobs are removed (Figure 13-6B). Set the external fine-tuning control to midrange. Then adjust the highest number channel fine-tuning screw for best picture and sound. Continue adjusting for best picture and sound on each successive channel that you receive. For instance, if channel 8 were the highest number channel, you would adjust that first, then channel 5, and finally channel 3. Always adjust from high to low. You shouldn't have to move the screws more than one or two turns. When no improvement is observed, be sure to reset the tuning screws to their original positions. A nonmetallic screwdriver is best to use for these adjustments. A 3/16" dowel rod shaped with a screwdriver point makes an excellent alignment tool.

13-7. Working with Picture Tubes

Does your TV screen take a long time to reach brightness? Do you have to pull the drapes and turn off the lights to see the picture clearly? Are there some brighter areas of the screen that show no definition? Do the contrast and brightness controls no longer have much effect? These are all signs of a picture tube wearing out especially if there are a lot of viewing hours on the tube.

The easiest and least expensive repair is to try a picture tube brightener. This is a little device that can be purchased in most electronics stores that fits between the base of the picture tube and its socket. In many cases it will improve the brightness considerably. When you purchase a picture tube brightener be sure to get the correct type. Here are some things to consider:

1. Black and white or color
2. Series or parallel heater connections (see section 13-3)
3. Type of picture tube base

Once you get the brightener, it is a simple task to install. Carefully remove the picture tube socket and insert it into the brightener. Then slip the brightener's socket onto the picture tube's base (Figure 13-7A). That's all there is to it!

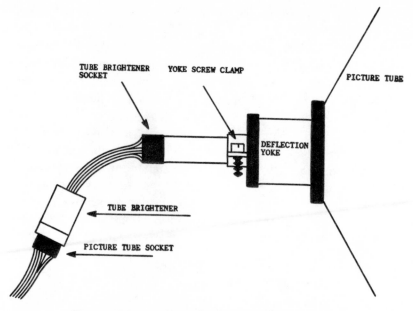

Figure 13-7A. Installing a picture tube brightener.

Occasionally the yoke around the neck of a picture tube will slip causing the raster to be in a slightly diagonal position (Figure 13-7B). This problem is a snap to fix. Figure 13-7A shows the position of the yoke around the neck of a picture tube. Inside the yoke are the two horizontal and two vertical deflection coils that shape the raster. Most yokes have a screw clamp to hold it in place. To readjust the yoke

Figure 13-7B. Diagonal raster caused by slipped yoke.

loosen the screw and turn the yoke assembly right or left in a circular manner to straighten out the raster. Then retighten the yoke clamp screw. *Caution:* The yoke has very high voltages present so be certain that you don't touch any bare wires or connections while making the adjustment.

Many yokes have centering adjustments mounted on them. Figure 13-7C shows the location of these adjustments. The yoke clamp does not have to be loose to make these adjustments. Simply slide the tabs on the centering mechanism around until the picture is centered. It's best to use a mirror to watch the picture tube screen when making these adjustments.

Some TVs have a width sleeve adjustment located on the picture tube neck under the yoke assembly. This is simply a thin metal wafer fastened to a piece of insulating material. By moving the metal wafer in its position relative to the yoke, the raster width can be varied. Loosen up the yoke clamp while pushing or pulling on the width sleeve to increase or decrease the width.

Black-and-white picture tubes are fairly easy to replace and require no special tools or alignment instructions. When handling picture tubes use extreme care not to force or strain any part of the tube, especially around the neck assembly. The air is evacuated from a picture tube causing a vacuum on the inside. Consequently there is a tremendous atmospheric pressure being exerted on the glass envelope. If a picture tube is dropped or bumped it could implode violently. This

CENTERING ADJUSTMENT TABS

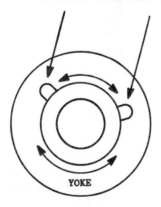

YOKE

Figure 13-7C. Centering adjustments on the yoke assembly.

simply means that as the glass structure cracks and breaks apart the pieces will be forced to the inside of the tube. As they collide they will then be thrown outward causing damage to anybody who is around.

To remove a black-and-white picture tube pull off the base socket. Next loosen up the yoke clamp and carefully slide the yoke off the neck of the picture tube. The only other connection is the high-voltage connection fastened to the side of the picture tube. Picture tubes can carry a high-voltage charge even when they have been off for a number of hours. You can safely discharge a picture tube by grounding one end of a clip lead to the TV chassis. Take the other end of the clip lead, clip it to the shaft of a screwdriver with a well-insulated handle and then touch the tip of the screwdriver to the high-voltage connection on the picture tube (Figure 13-7D). If the connection is insulated with a rubber cap, carefully insert the blade under the rubber boot until it

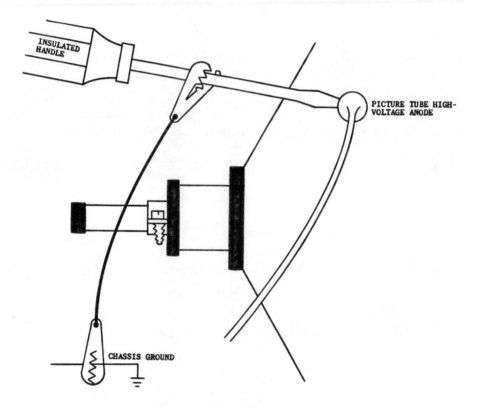

Figure 13-7D. Discharging a picture tube using a clip lead.

reaches the connection. The grounded screwdriver will immediately discharge any high-voltage buildup. After the high-voltage connection has been discharged, it is safe to remove the high-voltage lead.

Most high-voltage leads are fastened to the picture tube with a spring fit. Figure 13-7E shows the fastening arrangement. A spring clip fits into a recessed hole in the picture tube. When the two sides of the clip are squeezed together, it's a simple matter to remove it from the picture tube.

After the high-voltage lead is removed, inspect the mounting arrangement for the picture tube. Generally most picture tubes are fastened by a metal strap completely encircling the tube at its widest part. Sometimes the strap has to be removed; other times a few screws hold the strap and picture tube in the cabinet. It's best that the cabinet be placed so that the picture tube is resting on its face. In this position the tube will be safe and not drop out of place when the last hold-down connection is removed.

As soon as the old tube has been removed, keep it in the new tube carton. Fasten the mounting strap on the new tube in the exact position as on the old tube. Tighten it firmly but not excessively. Replace the picture tube, high-voltage lead, yoke, and tube socket. Apply power and you'll see the brightest picture you can imagine! You might have to make a yoke and centering adjustment to position the raster correctly.

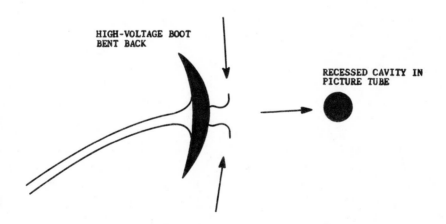

Figure 13-7E.   High-voltage lead connection to picture tube.

A color picture tube replacement is similar to the black-and-white, except that it's more involved. The color tubes have more assemblies around their necks. Also a multitude of controls should be readjusted when a color tube has been replaced. If you want to tackle a color picture tube replacement, you should have access to a dot and bar generator and a complete alignment procedure from the manufacturer for the picture tube and convergence controls.

INDEX

DC volts, checking: *(cont.)*
    electronic ignition, 148-149
    solenoid circuit, 160-161
    starting motor, 160-161
  reading a VOM, 33
  testing with neon lamp tester, 28-30
  thermocouple, gas furnace, 170-171
Deflection yoke, 196-197
Desk lamp:
  fluorescent, 52-53
  high-intensity, 53-56
  repairing, 46-50
Discharging, capacitor, 25-26
Dishwasher:
  operation, 84-85
  pump, 91-93
  solenoid water valve, 87-89
  timer, 85-87
Dispenser, solenoid, automatic washing
    machine, 100, 102
Distributor, automobile, 138-141
Double-insulated power tools, 76
Dryer controls, electric, 117-119
Dryer timer sequence chart, 112

E

Electrical appliances, small:
  checking line cord and plug, 74-76
  motor testing, 79-82
  theory, 73-74
  testing switch and controls, 76-78
Electrical code, 19
Electrical controls, air conditioner,
    126-128
Electrical plug, connecting line cord,
    46-50
Electrical safety:
  automobile battery, 17-18
  high voltage, 25-26
  house wiring, 18-25
  understanding shock, 16-17
Electric dishwasher repairs:
  operation, 84-85
  pump, 91-93
  solenoid water valve, 87-89
  timer, 85-87
Electric drill, cross-sectional view, 75
Electrician's knife, 37-38
Electric pump, dishwasher, 89-91
Electrodes, spark plug, 136-137

Electronic control unit, automobile,
    148-150
Electronic dryer controls, 117-119
Encapsulated control device, 77-78
Evaporator, air conditioner, 124-125

F

Fan-air condition switch, 127
Fan motor, air conditioning, 128-129
Feeler gauge:
  automobile:
    gapping spark plugs, 136-137
    setting point gap, 139
  gapping spark plug, 64-65
Field coils, universal motor, 79-82
Filing, spark plug electrodes, 136-137
Fine tuning, TV, 194-195
Fixture cord:
  connecting plugs, 46-50
  test lead construction, 31-32
Fluorescent lamps:
  ballast, 51-53
  instant start, 52-53
  repairing, 51-53
  starter, 51-52
Flywheel:
  adjusting air gap, 68
  construction and removal, 66-67
Four-way switch, checking with neon tes-
    ter, 44-45
Furnace, gas:
  blower control, 171-173
  blower motor, 173-175
  limit control, 171-173
  operation, 164-166
  thermocouple, 169-171
  thermostat, 166-169
Fusistor, 185-186

G

Gapping:
  magneto points, 66-67
  small engine coil, 68
  spark plug:
    automobile, 136-137
    small gasoline engine, 64-65

Gas furnace:
    blower control, 171-173
    blower motor, 173-175
    limit control, 171-173
    operation, 164-166
    thermocouple, 169-171
    thermostat, 166-169
Gasoline engines, small:
    ignition repairs, 62-71
        coil, 67-70
        condenser, 66-67
        magneto points, 66-67
        solid state, 70-71
        spark plugs, 63-66
        theory, 62-63
GFCI, 24-25
Graphite grease, 137
Ground fault circuit interruption, 24-25

H

Headlamp, automobile, repair, 57-60
Heater, checking with neon lamp tester, 29-30
Heating system, clothes dryer, 113-114
High-current batteries, 17-18, 145, 153-157
High-intensity lamp, repairing 53-56
High-tension cable, automobile, 140-141
High voltage safety, 25-26
High-voltage lead, TV:
    connection, 199
    discharging, 198-199
House wiring circuit:
    color code, 19
    safety, 18-25
Hydrometer, 153-156

I

Ignition system, automobile:
    coils, 141-144
    distributors, 138-141
    electronic ignition, 147-150
    emergency repairs, 144-146
    operation, 134-135
    spark plugs, 135-138
Ignition system repair:
    small gasoline engine, 62-71
        coil, 67-70
        condenser, 66-67

Ignition system repair: *(cont.)*
    small gasoline engine, *(cont.)*
        magneto points, 66-67
        solid state, 70-71
        spark plugs, 63-66
        theory, 62-63
Impeller:
    clothes dryer, 122
    dishwasher, 91-93
Incandescent lamp:
    high-intensity, 53-56
    repairing, 46-50
    three-way bulb construction, 50-51
Instant start, fluorescent lamp, 52-53
Instrument cluster removal, 60
Insulated alligator clips, construction, 31-32
Insulation, wire stripping, 37-39

J

Joints, solder, how to make, 35-36

L

Lamp, fluorescent, repairing, 51-53
Lamp socket-switch assembly:
    repairing and replacing, 49-50
    three-way, 50
Limit control, gas furnace, 171-173
Line cord:
    checking small appliance, 74-76
    connecting plugs, 46-50
    test lead construction, 31-32
Lubrication, universal motor, 79

M

Magneto points, checking and replacing, 66-67
Mercury switch, 166-167
Meter face, VOM, how to read, 32-34
Mica insulation, 80-82
Motor:
    automatic washing machine, 104-107
    blower, gas furnace, 173-175
    clothes dryer, 120-122
    disassembly, 79-80
    dishwasher, 91-93
    timer, dishwasher, 87

Turn signal wiring circuit, 59-60
Turret-type tuner, 193
TV:
  block diagram, 181
  controls, 186-189
  power sections, 178-179
  signal sections, 178-179
  trouble symptoms, 181-183
Two-speed motor, automatic washing machine, 104-107
Two-way valve, automatic washing machine suds return, 102

U

Universal motor repair:
  checking line cord and plug, 74-76
  motor testing, 79-82
  testing switch and controls, 76-78
  theory, 73-74

V

Vacuum tube testing, TV, 190-192
Variable-speed control, 77-78
Voltage relay, air conditioner, 127-128
Volt ohm milliameter, checking:
  ACV high-intensity lamp circuit, 53-54
  air conditioner compressor, 130
  automatic washing machine:
    solenoids, 101-102
    timers, 100
    water level pressure switch, 103
  automobile:
    cables and connections, 160-161
    coil, 142-143
    electronic ignition, 148-150
    solenoid circuit, 160-161
    starting motor, 160-161
  condenser, 66-67
  dishwasher:
    pump, 90-91
    solenoid, 88
    timer, 86-87
  dryer:
    heating elements, 114
    thermostatic controls, 116
    timers, 111

Volt ohm milliameter, checking: (cont.)
  gas furnace:
    blower, 172
    blower motor circuit, 174-175
    limit switch, 172
    thermocouple, 170-171
    thermostat, 168-169
  how to use, 32-35
  resistance spark plug cable, 140-141
  resistor spark plug, 137
  transformer resistance, 55-56
  tube filaments, 184-185
  washing machine motor, 106-107
VOM, see Volt ohm milliameter

W

Wafer-type tuner, 194-195
Wall fixture, checking with neon tester, 44-45
Wall outlets, checking:
  AC voltage, 130
  with neon lamp tester, 28-29
Wall receptacle, checking with VOM, 130
Washing machine, automatic:
  motor, 104-107
  operation, 96
  solenoids, 100-102
  timer, 96-100
  water level control, 102-104
Water control valve, solenoid, automatic washing machine, 100-102
Water level control, automatic washing machine, 102-104
Water valves, solenoid, dishwasher, 87-89
Width sleeve adjustment, TV, 197
Wire nut, 37, 45
Wire-stripping tool, 37-39
  electrician's knife, 37-38
Wiring circuit, turn signal, 59-60
Wiring diagram:
  automobile outside light, 56-60
  small gasoline engine, 63

Y

Yoke, 196-197